[英国]丽贝卡·阿诺德 著　朱俊霖 译

牛津通识读本·

时装
Fashion
A Very Short Introduction

译林出版社

图书在版编目(CIP)数据

时装 / (英)丽贝卡·阿诺德 (Rebecca Arnold) 著；朱俊霖译. —南京：译林出版社，2019.10
(牛津通识读本)
书名原文：Fashion: A Very Short Introduction
ISBN 978-7-5447-7588-5

I.①时… II.①丽… ②朱… III.①时装－研究－世界 IV.①TS941.7-9

中国版本图书馆CIP数据核字(2018)第270452号

Copyright © Rebecca Arnold, 2009
Fashion was originally published in English in 2009.
This Bilingual Edition is published by arrangement with Oxford University Press and is for sale in the People's Republic of China only, excluding Hong Kong SAR, Macau SAR and Taiwan, and may not be bought for export therefrom.
Chinese and English edition copyright © 2019 by Yilin Press, Ltd

著作权合同登记号　图字：10-2011-272号

时装　[英国]丽贝卡·阿诺德 ／著　朱俊霖／译

责任编辑　许　丹
装帧设计　景秋萍
校　　对　孙玉兰
责任印制　董　虎

原文出版　Oxford University Press, 2009
出版发行　译林出版社
地　　址　南京市湖南路1号A楼
邮　　箱　yilin@yilin.com
网　　址　www.yilin.com
市场热线　025-86633278
排　　版　南京展望文化发展有限公司
印　　刷　江苏凤凰通达印刷有限公司
开　　本　635毫米×889毫米　1/16
印　　张　19.75
插　　页　4
版　　次　2019年10月第1版　2019年10月第1次印刷
书　　号　ISBN 978-7-5447-7588-5
定　　价　39.00元

版权所有　·　侵权必究

译林版图书若有印装错误可向出版社调换。质量热线：025-83658316

序　言

臧迎春

给这本书作序，似乎是一种缘分。

展览"恶毒的缪斯"在伦敦的维多利亚和阿尔伯特博物馆，以"魅影：时装回眸"的主题展出时，我恰在伦敦，这个展览令我沉浸其中，印象深刻。而2004年赴伦敦中央圣马丁艺术设计学院做博士研究之初，我在查令十字街校区旁的书店发现了一本书，并用了整整一个月的时间去读，这本书正是时装史学家、理论家卡罗琳·埃文斯发表于2003年的著作《前卫时装：壮美、现代与死寂》，她对时装及时装史的重要解读对于我后来的研究产生了深刻的影响。

上述展览和著作在丽贝卡·阿诺德《时装》一书的开篇都被重点提及，让我顿生老友重逢的久违之感。

对于时装的认识，至今仍存在很多的偏见。有人认为，时装就是流行的、好看的衣服；有人认为，时装是身份、财富的标志；还有人认为，时装是一种扮装拟态的道具。时装是什么？从概念上讲，它是在一段时间内、在比较广阔的地域范围内，很多人都穿着的流行服装。时装看起来是具有很强的时间性和地域性的,但实际上时装的根来自深厚的历史,而它的触角又不断探向遥远的未来。正如丽贝卡·阿诺德在《时装》一书中所作的较为透彻的阐释:"时装总是追赶着最新的潮流,可与此同时它又一刻不停地回首过去。"(引言第4页)

时装之所以备受争议,是因为它本身就是一个矛盾的集合体。既可以是阳春白雪,又可以是下里巴人;既可以全球化,又可以本土化;既可以强调学术性,又可以强调商业性;既可以成为功能性强的装备,又可以成为非功能性的装饰。人们对于时装看不完整,说不清楚,既爱又恨,又难舍难分。

无疑,时装与社会、时代,与文化、科技、经济、政治甚至是军事都密切相连,是多学科交叉的一个复杂系统,要解释清楚谈何容易?每个学者的研究往往都是从某一个角度切入,进而力图阐述整个系统的状态和运作方式。在这本书里,作者从"探寻时装作为一个产业的运作方式,以及它如何连接起更广泛意义上的文化、社会及经济议题"的角度入手,这是一个不错的切入点。文中指出:"自1960年代开始时装成为一门可以进行严肃学术讨论的学科,由此促进了对其作为图像、客体及文本的诸多分析。"(引言第7页)其实从那时起,人们就开始以严谨的学术态度从一系列重要角度来审视时装。"时装研究天然的跨学科特性折射出它与历史、社会、政治和经济等诸多背景的紧密关联,也与诸如性别、性向、民族和阶级等更加具体的问题联系密切。"(同上)我们在半个多世纪的研究中可以发现,"时装"是最"瞻前顾后"的学科。它一方面最敏锐地关注并反映着社会发展的前沿动态,另一方面又不断从历史中汲取灵感和丰富的营养。

它在思想层面和实践层面上都积累了极其丰厚的成果。诸多设计师也在这一领域进行了积极的探索，并进一步印证了时装的这一特性。

从"款式商人"罗斯·贝尔坦，到可可·香奈儿，再到卡尔·拉格斐，他们都是深谙"创新"与"对于历史与时代热点绝妙借用"的高手。19世纪中叶，沃思等人开始以"时装与服饰设计师"的名头宣传自己，并将自己刊载在当时巴黎黄页中"工业设计师"的目录中，这开启了现代时装设计师的历史。保罗·普瓦雷可以称得上是第一位现代时装设计师，他"汲取了从现代主义到'俄罗斯芭蕾'等各种当代艺术与设计灵感"，其"高级时装形象的格调又通过他自己的香水产品销售传播得更为深远"（第11页）。二十世纪二三十年代的维奥内夫人、夏帕瑞丽，四五十年代的迪奥、巴伦西亚加，六十年代的安德莱，七十年代的韦斯特伍德，八十年代的川久保玲、山本耀司等，此后的众多设计师对应不同的时代需求，各显才能，将高级时装"推到了巨大的全球奢侈品市场的首位"。也许很多人无法理解，高级时装本身并不能盈利，为何还有存在的价值？本书作了很好的解答："即便这些孤品式的设计本身并不能创造什么利润，但它们带来的大量关注巩固着处于高级时装产业核心位置的设计师们始终如一的重要性。"（第13页）150多年来，现代时装发展的历史告诉我们，设计师的创新是时装发展的灵魂，它在不断拓展着时装的边界，丰富着时装的内涵。

在本书当中，丽贝卡不仅探讨了时装与设计师的关系，还探讨了时装与个性、传播、艺术、商业、全球化等方面的关系。

时装设计师的创新是围绕个性展开的,因为时装关注的核心是个性。"通常认为时装是从文艺复兴时期兴起的,它是商贸活动、金融行业的发展,人文主义思潮所激发的对个性的关注,以及社会阶级结构转变等多方因素共同作用的结果,其中阶级转变使得人们渴望视觉上的自我展示,并使更多的人群能够实现这一愿望。"(引言第8—9页)其实,这种需求一直存在着,"当城市的巨大发展带来更多的个性泯灭,时装成为能够构建身份,并让社会、文化及经济地位一目了然的重要途径"(第8页)。从这个意义上讲,时装本身就是一种个性、身份识别的标志。

同时,时装也是一种交流方式。"时装信息通过雕版绘画、行脚商贩、书信往来,还有17世纪末发展起来的时尚杂志得到不断传播,这让时装越来越可视化,人们也越来越渴望时装。"(引言第9页)"时装不只是服饰,也不只是一系列形象。实际上,它是视觉与物质文化的一种生动体现,在社会和文化生活中扮演着重要的角色。"(引言第9—10页)时尚无边界,"时尚媒体、摄影的发展,以及19世纪末出现的电影,它们使时尚形象得到了空前的广泛传播,且激发了女性对更为丰富多样及更迭迅速的时装款式的欲望"(第8页)。 时尚的传播不仅仅是信息的扩散,更重要的是,它成为人与人、人与社会之间的最为直接的交流手段,它对于自身形象和身份的塑造功能,使它成为每个人的社会标签和商业包装。

很多人对时尚嗤之以鼻是因为"时尚固有的短暂易逝而又物质至上的缺陷"。而安迪·沃霍尔意识到,"时尚、艺术、音乐与流行文化之间已经结盟。将先锋流行音乐与基于色彩明艳的金属、塑料以及撞色印花的抛弃型和实验性服装进行融合,不仅表达出这一时代的创造激情,还帮助确立了时尚标准"(第26页)。沃霍尔因此成为将艺术、时尚和商业完美结合的又一个典范。

我颇为认同"有时人们将时装与艺术相提并论,以赋予它更高的可信度、深度以及意图。不过,比起对时装真实意义的揭示,这种做法也许更多地暴露了西方世界对于时装缺乏这些特质的担忧"(第29页)。实际上,"它并不需要被冠以艺术之名来佐证其地位",它本身的价值绝非艺术所能替代。我更加认可俄国构成主义派设计师瓦瓦拉·史蒂潘诺娃的观点:"认为时装将会被淘汰或者认为它是一种可有可无的经济附属品其实大错特错。时装以一种人们完全可以理解的方式,呈现出主导某个特定时期的一套复杂的线条与形式——这就是整个时代的外部属性。"(第40页)可惜的是,许多人因为不能解读或者无从解读,从而无法深入理解时装,并且因为无知而轻视时装。

"作为视觉文化的两个重要组成部分,时装与艺术始终表达着并且不断构建起诸如身体、美还有身份等各类观念。"(第44页)"理解时装应当以它自己的语境为基础,这使得时装与艺术、文化其他方面的相互交织更加妙趣横生。它为艺术、设计以及商业在一些时装实践者的作品中发生联系、交织重叠开辟了新的路径。的确,时装能够如此令人着迷而又令某些人感觉难以捉摸的一个重要原因就是,时装总是能侵吞、重组并挑战种种既有定义的边界。"(第29—30页)时装的这一特点为当代视觉文化提供了无限可能,也成为当代视觉艺术的试验场。

作为与商业的重要关联,本书多次提到成衣的发展。"1966年,伊夫·圣罗兰的'塞纳左岸'(Rive Gauche)精品店开业,以长裤套装与色彩灵动的分体款时装来响应流行文化并认同女性社会角色的转变。"（第14页）20世纪是个时装业高速发展的时期，1960年代"一般被视为大规模生产、年轻化的成衣开始以前所未有的姿态引领时尚的关键节点。美国的邦妮·卡欣等设计师，英国的玛丽·奎恩特等人，还有意大利包括璞琪在内的设计师们当时都在市场的各个层次里确立着自己的时尚影响力，塑造了时装设计、销售以及穿着的新范式"（第15页）。1980年代"安特卫普与东京展示了各自培养出众设计师的实力，一批设计师开始成名，其中就有比利时的安·迪穆拉米斯特和德赖斯·范诺顿，以及日本籍的Comme des Garçons品牌设计师川久保玲和山本耀司。到了21世纪初期，中国和印度也像其他国家一样，开始投资自己的时装产业，并培养自己的季节性时装秀"（第16页）。

在丽贝卡的描述中，一切似乎顺理成章。但在中国，现代时装的发展之路却格外不寻常。

20世纪初，西风东渐，以上海为代表的一些城市的时装逐渐吸取西方服饰所特有的"构筑性"造型方式，在裁剪上产生显著变化。中国的时装开始从"直线裁剪"向"曲线裁剪"转变，其几千年的"宽衣文化体系"受到"窄衣文化体系"的强烈冲击。这是东西方哲学、美学体系的一次剧烈的碰撞与交锋，而交锋的地域就是中国。

1920年代以后，西方时尚体系的影响在中国得到进一步深化，即使是在战争时期，这种影响也并没有完全中断。改良式旗袍、中山装、学生装、布拉吉等都是这一时期的代表性时装。20

世纪中期以后，直到改革开放以前，中国的服装已经在主体上转变为深受西方影响的"窄衣"——分部件曲线裁剪，再组装在一起的适体服装，不再是传统美学思想下的，整体的、直线裁剪的、并不完全适体的"宽衣"。1970年代末的改革开放，让中国成为全球时尚产业中的重要一环，经过40年的发展，中国的时装在很多一线城市已经与国际同步。我们在看到香奈儿、迪奥、阿玛尼等一线品牌出现在北上广深等一线大城市的同时，也可以看到ZARA、优衣库、H&M等快时尚品牌充斥在各个城市的核心街区和重要商业网络平台上。

本书提到，达娜·托马斯在《奢华：奢侈品为何黯然失色》一书中引述了福特的评论："本世纪属于新兴市场……我们（的事业）在西方已经到头了——我们的时代来过了也结束了。现在一切都关乎中国、印度和俄罗斯。这将是那些历史上崇尚奢华却很久没有过这种体验的众多文化重新觉醒的开始。"（第83页）一方面，我们看到20世纪末以来，中国市场确实出现过追逐名牌，追求奢侈的潮流；而另一方面，我们也清楚地意识到，当今时代基于可持续发展的要求，设计道德和设计责任被提到越来越重要的地位。极简主义的生活方式呈现越来越强劲的势头。在21世纪，中国时装以一种更加混杂、多元化的状态存在。这与多元化的生活方式，社交平台、网络等多种传播方式，线上、线下等多种营销方式，以及后工业时代、工业时代和前工业时代等不同地域的发展不平衡都有密切关联，是其他国家在时装发展过程中都没有遇到过的复杂状态。

有一些学者认为"时装终结了"。本书认为："社会、文化以及政治生活方式与态度的这些不同方面，逐渐与时装的诞生、传播以及越来越全球化的特性联系在一起。时装因而并未终结，但它确实发生了变化，并且极有可能处在另一次深刻变革的边缘。随着非西方时装体系暗自发展壮大，经济衰退又席卷而来，时装主力很可能会转向东方。"（第130页）但笔者认为，时装中心的转移并不简单，除了经济、文化、政治等诸多因素以外，国际时装产业在世界范围内的上下游关系，或者说是现有的时装体系，在短期内是很难打破的。重建国际时装体系需要时间，更需要明确的思路。因此对于中国来讲，对于时装全面的认识很重要，充裕的资金很重要，产业基础很重要，成熟的传播与宽松的市场环境很重要，但重中之重，人才培养是关键。

值得一提的是，中国的时装设计教育是与改革开放同步的。1980年中央工艺美术学院（1999年并入清华大学）设立了服装设计专业，经过近40年的发展，逐渐建立了系统的服装设计教育体系。其毕业生遍及全国，逐渐在国际上形成越来越大的影响力。国内其他院校也纷纷设立服装与服饰设计或服装工程专业，从不同层面培养了大量人才。与此同时，又有大批学生留学国外，在不同文化和教学体系下学习，融合东西方的文化，了解全球化市场下国际时装体系的运作方式，这都为中国时装的发展奠定了基础。

但是，优秀时装设计师在中国的发展也还面临着许多困难，这与其他国家的情形相似。"如果设计师在职业生涯之初就过快地斩获盛名，而此时的他们还没有找到有力的资金支持，不具备与订单需求相匹配的生产能力，他们的事业便难以成长。尽管如此，媒体报道仍然被认为对于树立品牌，并最终找到来源可靠的资金投资至关重要。"（第55—56页）因此，时装设计师的孵化器、成熟的时尚评论对于中国设计师的成长具有重要意义,它需要来自政府、产业、媒体和院校的综合支持。

"时装"是个难以讲述的话题,而丽贝卡·阿诺德的《时装》一书不仅系统、深刻地阐述了一百多年来欧洲时装形成的体系、发展的脉络,探讨了时装与艺术、商业之间的复杂关系,而且也洞悉了时装在当代所面临的全球化问题、道德问题和可持续发展问题,其思想敏锐透彻,文字鞭辟入里,是值得一读的好书。

丽贝卡在文中指出:"时装业包含了一系列相互交融的产业领域,这个领域的一端聚焦生产制造,另一端关注最新潮流的推进与传播。"(第47页)时装从来都在"路上",一直在不断变革。近些年来"可持续发展"成为时装领域越来越重要的主题:"旨在管制服装并创造出不伤害动物、人和环境的服装的推动力在20世纪晚期和21世纪初的形式开始逐渐进入主流观念,同时也融入商业时装之中。"(第101页)而想要实现让时装更符合道德准则的目标,"只有靠大规模重组社会与文化价值观,并变革全球化产业模式"才能够实现。由此看来,时装领域的变革在即,时装之路任重道远。

在全球化过程中,时装的同质化问题令人担忧。而"塞内加尔现象"具有积极意义。首都达喀尔的设计师"创造出以当下流行的本土风格、传统染色及装饰元素、国际名流,还有法国高级时装为灵感的服装。全球化的贸易网络使得塞内加尔商人能够订购北欧的纺织品设计,收购尼日利亚的织物,然后在欧洲、美洲和中东开展贸易活动。整个国家的时装体系因而整合了本土与全球的潮流,创造出最终到达消费者手里的时装。它很快成为全球化时装产业的一部分，但同时又保留着自身的商业模式与审美品味。达喀尔这座充满活力的时尚都会是 21 世纪各国时装产业能够共存、共生的典范"（第121页）。这样一种模式为不同地域既保持本土文化特色，又有效利用全球化系统提供了丰富资源，同时为丰富发展多样化的时装系统做出了积极探索。

21 世纪，新技术、新材料不断涌现，大批时装专业人才得到系统培养，中国等发展中国家的财富与工业生产能力不断提升，时尚潮流与生活方式通过互联网等新的媒介和平台快速、大面积传播，越来越多的人们开始质疑原有的、已经固化的时装产业系统，期待建立一个更加符合可持续发展的、更好地服务于美好生活的新时装体系。中国作为具有深厚文化艺术底蕴、多种多样手工艺传统的国家，加之全球最活跃的新兴经济市场和大批设计、产业人才的助力，必将成为国际时装体系转型发展的重要力量。

2019年7月写于清华园

献给阿德里安

目 录

致谢　1

引言　3

第一章　设计师　1

第二章　艺术　24

第三章　产业　45

第四章　购物　65

第五章　道德　85

第六章　全球化　106

结语　126

索引　131

英文原文　143

致　谢

在这里我想要感谢安德烈娅·基根,她是牛津大学出版社负责此书的编辑,感谢她对于这本书全力的支持与始终的鼓励。我还要感谢伦敦皇家艺术学院设计史系的所有同仁与同学们。卡罗琳·埃文斯提出的精彩建议,夏洛特·阿什比与比阿特丽斯·贝伦对书稿所做的细致点评,都使我感激不尽。感谢艾利森·托普利斯、朱迪思·克拉克还有伊丽莎白·柯里给这本书的有益建言。最后,衷心地感谢我的家人,感谢阿德里安·加维所付出的一切。

引 言

"恶毒的缪斯"(*Malign Muses*)是朱迪思·克拉克2005年在安特卫普时尚博物馆策划的一场开创性大展,展览集合了最新款时装与古董服装,并将它们分列于一系列令人叹为观止的布景之中。布景设计看上去与19世纪的露天市场类似,简洁单调的木质结构组成了可以转动的展架,鲁本·托莱多创作的大型黑白调时装绘画作品更为展览增添了一丝略带魔幻又戏剧化的感觉。这次展览重点呈现了时装的令人兴奋之处与壮观场面。约翰·加利亚诺和亚历山大·麦昆复杂精妙的时装设计,与两次世界大战之间的高级定制时装杂陈一堂,其中就有艾尔莎·夏帕瑞丽著名的"骨骼裙",这袭黑色的紧身裙上装饰着突起的骨骼结构。一条造型夸张的1950年代克里斯汀·迪奥晚礼服,丝绸用料极具质感,上身是有型的胸衣,下身是拖曳的长裙,腰身之后打上了蝴蝶结;与它陈列在一起展示的,是一件19世纪晚期印度生产的精致白色棉布夏装,装饰有印度传统的链式针法刺绣纹样。比利时设计师德赖斯·范诺顿1990年代末制

图1　2005年由朱迪思·克拉克设计、策展的安特卫普时尚博物馆"恶毒的缪斯"展览一角

作的宝石色印花与光洁的亮片设计立在一套色彩艳丽的克里斯汀·拉克鲁瓦1980年代套装旁边。各式服装复杂精美的组合在克拉克设计巧妙的布景烘托之下使人一目了然，她的排布方式集中于时装对历史进行借鉴与参考的种种丰富多变的形式之上。这场展览的戏剧式场景吸收了18世纪的即兴喜剧和假面舞会元素，也直接借鉴了当代时装设计师在他们每一季时装秀场中对戏剧感与视觉冲击力的运用。

"恶毒的缪斯"后来移师伦敦的维多利亚和阿尔伯特博物馆，在那里被重新命名为"魅影：时装回眸"（*Spectres: When Fashion Turns Back*）。这个新名字传达出了时装最为核心的一个矛盾：时装总是追赶着最新的潮流，可与此同时它又一刻不停地回首过去。克拉克极为有效地呈现了这一核心的对立，鼓励参观者思考时装丰富的历史,同时把它与时装的最新议题联系起来。通过把时代不同,但技法、设计或是主题相似的服饰并列陈设,克拉克实现了这一效果。展览的成功也要归功于克拉克与时装史学家、理论家卡罗琳·埃文斯的密切合作。通过借用埃文斯在她2003年的著作《前卫时装:壮美、现代与死寂》(*Fashion at the Edge: Spectacle, Modernity and Deathliness*)中对时装及时装史的重要解读,克拉克揭示了时装背后鲜为人知的推动力量。埃文斯向世人清楚展示了旧时代是如何一直影响着时装的,正如它在始终影响着更广泛意义上的文化一样。对过去的借鉴可以增加新式夸张设计的可接受度,并且将其与曾经备受崇敬的典范联系在一起。这在展览中格雷夫人(Mme Grès)裙装设计的纤巧褶裥上就能看出来,它正是以经典的古代服饰为灵感来源的。时装始终追求着年轻和新鲜,它甚至可以表达人类对死亡的恐惧,荷兰的维果罗夫品牌(Viktor and Rolf)就用全黑的哥特风格礼服清晰地传递着这些信息。

参观者们因此不仅可以看到时装在视觉与材料层面对其历史的运用,又能够通过一系列滑稽的表演片段,得以探究服装更深层的意义。作为对展览露天集市主题的延续,一系列巧妙构思的视觉错觉借用镜子迷惑着参观者的双眼。展览中的裙装看上去一会儿出现一会儿又消失,它们要从小孔中窥视,或者被放大或缩小。于是,参观者们不得不全神贯注地观察他们眼中的一切,又不断怀疑自己对眼中事物的判断。

展览引发了参观者对时装含义的思考。时装与服装形成了鲜明比照,后者通常被人们视为一种更加常态化、功能性的衣着形式,其改变是非常缓慢的,而时装则立足于新奇与变化。它那周期性逐季改变款式的特性在托莱多的圆形绘画作品中得到了体现，画中呈现了一个永不停歇的时装剪影环，每个剪影都与下一个不同。时装也常常被人们认为是一种加之于服装的"价值"，让消费者们渴望拥有它们。展览华美而戏剧化的布景反映了时装秀、广告以及时装大片通过展示理想化的服装形态来诱惑、吸引消费者的各种手段。同样地，时装也可以被看作是同质化的一种形式，它煽动所有人都以某种特定方式穿着打扮，可是与此同时，它又追求个性与自我表达。20世纪中叶高级定制时装在时装业的独断专行，以迪奥的时装为例，与1990年代时装的丰富多样形成对比，从而强调了这种矛盾。

这种矛盾引导着参观者去理解可以在任何时代存在的各种类型的时装。甚至在迪奥的全盛时期，仍然有不同的时尚服饰选择，不管是加利福尼亚设计师们简洁的成衣风格，还是"不良少年"（Teddy boys）的反叛时装。时装可以生发于不同的源头，可以经设计师和杂志之手打造出来，也可以在街头环境中有机地演化。因此，"恶毒的缪斯"展览本身也成了时装史上一个意义非凡的节点。它把旧时代与新时代时装中看似毫不相干的元素统一到了一起，通过感官的布展方式呈现，令观众倍感愉悦又沉迷其间，但又引导他们明白时装的意义远超其表象。

正如展览所揭示的，时装总是建立在矛盾之上。对某些人来说，它们是曲高和寡的精英，是高级定制工艺与高端零售商的奢华天地。对另一些人而言，它们则不断更新，随手可抛，在随便一条高街都能买到。伴随新兴"时装都会"的逐年发展，时装越来越全球化，同时它又可以非常本土化，形成某个小群体专属的小规模时装风格。它可以纳入专业的学术著作或是闻名的博物馆,也可以出现在电视上的形象改造节目与专题网站里。正是时装这种模棱两可之处让它如此令人着迷,当然也引得人们冷眼相待,嗤之以鼻。

时尚风潮(fashions)可以诞生在各个领域,从学术理论到家具设计甚至是舞蹈风格。不过,就一般意义而言,特别是在这个词以单数形式出现时,它指的是穿衣的时尚。在这本《时装》里,我将探寻时装作为一个产业的运作方式,以及它如何连接起更广泛意义上的文化、社会及经济议题。自1960年代开始时装成为一门可以进行严肃学术讨论的学科,由此促进了对其作为图像、客体及文本的诸多分析。从那时起,人们就从一系列重要角度来审视时装。时装研究天然的跨学科特性折射出它与历史、社会、政治和经济等诸多背景的紧密关联,也与诸如性别、性向、民族和阶级等更加具体的问题联系密切。

罗兰·巴特在其符号学著作《流行体系》(1967)与《时尚语言》中从意象与文本的相互作用出发对时装进行研究,后者辑录了他1956年至1967年间的文章。从1970年代开始,文化研究成为探求时装与身份的新平台:例如,迪克·赫伯迪格在1979年的文本《亚文化:风格的意义》中,说明了街头时装是如何在青年文化的影响下演化的。1985年,伊丽莎白·威尔逊所著的《梦中装束:时尚与现代性》一书从女性视角对时装在文化和社会上的重要性下了一个重要的断言。艺术史一直以来都是一种重要的方法,它能够细致地分析时装与视觉文化相互交织的不同形式,安妮·霍兰德与艾琳·里贝罗进行的研究就是个很好的例子。珍妮特·阿诺德等人采用了一种基于博物馆研究的方法,她通过观察博物馆收藏的时装,细致地研究了服装的剪裁与结构。多样化的历史研究法对研究时装产业的本质及其与特定背景议题的关系十分重要。这一领域里有贝弗利·勒米尔基于商业角度的研究，也包括我本人的研究，还有克里斯托弗·布鲁沃德与文化史相关的研究。自1990年代以来，社会科学界的学者们开始对时装极为感兴趣：丹尼尔·米勒和乔安妮·恩特威斯尔两人的研究成果就是这股研究趋势中非常重要的代表。卡罗琳·埃文斯令人印象深刻的跨学科研究交叉借鉴了各家之法，成果卓著。专业院校里的时装研究同样异彩纷呈。艺术院校极为重视时装研究，将其作为设计专业课程中的学术培养内容，但它已经扩展到从艺术史到人类学等院系之中，同时也成为本科与研究生阶段的专修课程。

学术界对时装的兴趣一路延伸到了收藏有重要时装藏品的众多博物馆中，包括悉尼动力博物馆、纽约大都会艺术博物馆服饰馆以及京都博物馆等。策展人对时装的研究催生了大量的重量级展览，数目巨大的观展人群充分说明了人们对时装的普遍关注。特别是，展览在策展人的专业知识与当前的学术观点之间，在时装的核心即展出的服饰本身与帮助创造我们心中时装概念的种种图像之间，建立了清晰易懂的联系。

自文艺复兴以降，至今已发展出一个庞大的、国际化的时装产业。通常认为时装是从文艺复兴时期兴起的，它是商贸活动、金融行业的发展，人文主义思潮所激发的对个性的关注，以及社会阶级结构转变等多方因素共同作用的结果，其中阶级转变使得人们渴望视觉上的自我展示，并使更多的人群能够实现这一愿望。时装信息通过雕版绘画、行脚商贩、书信往来，还有17世纪末发展起来的时尚杂志得到不断传播，这让时装越来越可视化，人们也越来越渴望时装。随着时装体系的发展，它逐渐吸纳了学徒制和后来的院校课程，以此培育新的设计师与工匠，此外还有手工以及后来工厂化的纺织品与时装生产、零售行业，以及从广告到造型和时装秀制作等丰富多样的营销产业。时装的发展从18世纪晚期开始加快了脚步，等到工业革命正值顶峰的19世纪后半叶，时装已经涵盖了许多不同类型的流行风格。这一时期，为不同客户单独量体裁衣的高级时装作为一种精英化的时装形式在法国逐渐形成。将设计师的想法明确化的高级时装师们不只是这些手工服饰的创造者，更是不同时代时尚理念的制造者。早期重要的高级时装师，比如露西尔，将自己精心设计的时装用专业的模特展示出来，探索了借助时装秀为自己的店带来更多知名度的可行性。露西尔也看到了另一股重要的时装趋势，即不断增长的成衣贸易，它能够快速且轻松地生产大量服装，并将它们推向更广泛的受众。露西尔造访了美国，在那里销售自己的设计，甚至撰写了流行时装专栏，这些都突显了高级时装风格与流行的成品服装的发展之间千丝万缕的关联。尽管巴黎主宰着高级时装的种种典范，但世界各地的城市仍打造着自己的设计师与时装风格。到20世纪晚期，时装真正地全球化起来，出现了像埃斯普利特（Esprit）和博柏利（Burberry）这样的品牌巨头，产品销售遍布全球，发源于西方世界之外的时装也得到了更多的认可。

时装不只是服饰，也不只是一系列形象。实际上，它是视觉与物质文化的一种生动体现，在社会和文化生活中扮演着重要的角色。它是一股庞大的经济驱动力，位列发展中国家前十大产业之一。它塑造着我们的身体，塑造着我们审视别人身体的方式。它能够让我们以创作自由恣意表达另类的身份,也可以支配人们对美丽和可接受的定义。它提出了重要的伦理与道德质疑,联结起殿堂艺术与流行文化。虽然这本《时装》主要关注主导时装设计领域的女装,它仍分析了许多重要的男装案例。它将聚焦时装发展后期的几个阶段,同时也会回溯19世纪之前的重要先驱,以此展现时装是如何演化至今的。书中将会探讨主导时装产业的西方时装,但同样会对这种支配地位进行质疑,并展示其他时装体系是如何发展并与西方时装交叠的。我还将向读者们介绍那些与时装产业相互连接的领域,呈现时装是如何被设计、制造以及销售的,并剖析时装与我们的社会文化生活之间重要的联系方式。

第一章

设计师

香奈儿2008春夏高级定制时装秀上,一件巨大的品牌标志性开襟毛呢外套模型矗立在秀场中心的旋转平台之上。这件纪念碑似的"外套"以木头制成,配以水泥灰色调的喷涂,在模特们身后高耸着。模特们从"外套"前襟敞开的部分鱼贯而出,昂首阔步地在一众时尚媒体、买手和各界名流前走过,来到品牌的交扣双"C"标志前停步亮相,而后消失在可可·香奈儿留下的这个符号性标志之中。模特们身着的服装色调简洁,再一次反映出品牌的传统:生动的黑白二色搭配鸽子灰与极浅的粉色。这一系列的服装自品牌的斜纹软呢开襟外套衍生出来,整个香奈儿始终在真正意义或隐喻意义上被这款设计影响着。而这次品牌以当代手法重新打造了这一经典款式,使它显得轻盈而有女人味,褶边处分离出一簇簇叶片装饰,或是在修身剪裁上遍布亮片,下身着略有曲线的半身裙,其精致廓形源自海贝的天然形态。秀场的布置与展示的服装都是品牌渊源的缩影,体现在它们将可可·香奈儿酷爱的优雅的半裙套装、闪亮的人造珠饰以

图2 卡尔·拉格斐设计的2008版经典香奈儿套装

及层叠的晚礼服所进行的组合之上，同时又融合了品牌目前的设计师卡尔·拉格斐对当代的敏锐见解。

香奈儿发展成为20世纪最为知名和最具影响力的高级时装店之一，突显了成功时装设计的许多关键元素，显示了设计、文化、商业与至关重要的个性这四者间的相互关系。可可·香奈儿作为社会和时尚版面的显赫人物在1910年代到1920年代间的崛起，从夜场歌手到高级时装师的神话般的发迹，还有关于她众多情人的绯闻，这些都为她那简洁现代的设计风格赋予了一种刺激且神秘的气息。她的设计本身就耐人寻味，体现了对线条明快而层次简单的日装，以及更女性化且戏剧化的晚装的当代时装追求。她认为女性应该穿着简洁，要像她们身着小黑裙的女仆们一般，当然，克洛德·巴扬引用香奈儿的这句话是想提醒女性"简单不代表贫穷"。她钟情于混搭天然与人造珠宝，不断借鉴男性着装，这令她蜚声国际。可可·香奈儿的传记为她带来了公众的注意与好奇心，这对于提高她的时装店的知名度是极为必要的，也让她作为一名设计师和风云人物而引人注目。尤为重要的是，她对品牌进行多元经营，发展出配饰、珠宝以及香水等品类，同时还将自己的设计销售给美国买手，这把她的时尚精髓扩展到了高级时装消费人群之外的一个空前广阔的市场，确保了她在经济上的成功。

1980年代，时装评论家欧内斯廷·卡特总结了香奈儿的成功，认为其建立在"个人魅力的魔法"之上。与可可·香奈儿毋庸置疑的设计与造型能力同样重要的，是她对于理想化自我形象的营销能力，以及能够代表自己无可挑剔的顾客形象的能力，这使得她的品牌如此富有吸引力。香奈儿设计了自己的形象，而后将这一形象兜售给整个世界。许多后来者跟随了她的脚步：从1980年代开始，美国设计师唐娜·凯伦成功地为自己打造出一个不断奋斗的母亲与女商人形象，专为那些像她一样的女性设计服装。与她相反，多纳泰拉·范思哲则始终脚踩高跟鞋、身着极富魅力的紧身服装出现在镜头前，她富豪般的生活方式同样反映在范思哲（Versace）品牌那些珠光宝气的奢华时装中。

香奈儿现任设计师卡尔·拉格斐代表了这一命题下另一种不同的表现形式，比起呈现他的消费者们的生活方式，他的个人风格彰显了他作为造诣极深的美学大师的地位。如果说香奈儿是她的追随者们的时尚偶像，代表了一种20世纪初期时髦优雅、流线型灵动女性特质的现代主义理念，那么拉格斐则是为现如今的时代重新建构起来的一个帝王般的时髦标杆。他的个人风格中的关键元素在他于香奈儿麾下工作期间始终如一：深色套装，长发梳向脑后扎起马尾，偶尔会上粉涂成白色。再加上他不时喜欢拿在手里不断扇动的黑色折扇，他的形象令人不禁想起古代的法兰西君王。这些都彰显了高级时装的精英地位，以及香奈儿风格的延续统一，而他不断投身各种各样的艺术及流行文化事业又维持了他走在时尚前沿的公众形象。

香奈儿于1971年辞世，品牌的金字招牌也随之而去，其销售业绩与时尚公信力均开始衰减。在拉格斐的双手操持下，香奈儿品牌得以重振。自1983年加入品牌以后，他为品牌设计了高级时装系列、成衣系列以及配饰产品，较好地平衡了对于品牌特色同一性的需求，以及同等重要的能够反映并预见女性穿着想法的愿望。拉格斐在自由职业阶段为包括蔻依（Chloé）、芬迪

（Fendi）等不同成衣品牌工作的丰富经验，充分证明了他的设计实力以及能够创造出足以掀起新潮流、同时修饰美化女性身形的服装这一至关重要的能力。他通过融合经典与流行文化元素来维持香奈儿的影响力，并重振其时尚地位。他的2008春夏高定系列充分体现了这一点，也显示了他的商业头脑。尽管他的年龄在不断增加，品牌的忠实顾客们脑中想起的又始终是他基于经典开襟毛呢外套的各种改良造型，但这一系列的色调是年轻的，灰暗的颜色会配上充满少女感的荷叶边和轻柔织物配饰进行中和。拉格斐因而着眼未来以保香奈儿屹立不倒，并始终鼓动着新的年轻顾客穿上这一标杆品牌。

高级时装设计师的演化

回顾过去的历史，绝大多数服装是家庭自制的，或者是从几间铺子买来织物或辅料后再由本地裁缝和制衣师制作。到了17世纪末，一些裁缝，尤其在伦敦萨尔维街一带，开始以技艺最为纯熟、款式最为时髦而闻名，人们会从其他国家赶来在亨利·普尔等比较知名的几家定做西装。尽管某些裁缝公司确实在特定时期代表着时髦的风格，但男装设计师们在20世纪下半叶之前始终未能企及与其同业的女装设计师们的地位与名望。"裁缝"这个字眼意味着协同操作，不管是参与西服制作不同环节的各个工匠之间，还是与顾客就面料、风格及剪裁方案选择所进行的深入讨论。与此相反，到了18世纪晚期，女性时装的开拓者们则开始衍化出自己个性化的气质。这反映了女装中的创意与奇想有着更宽广的空间，它还取决于贵族阶层时尚引领者们与她们的制衣人之间逐渐形成的不同关系。尽管哪怕最知名的裁缝师也要与他的顾客非常紧密地就服装风格进行沟通，女装制衣师们却开始强力推出自己的风格。

尽管时装就参与其制作的人数而言一直保持着基本的协同工作流程，但它开始与个人的设计实力和时尚洞察力密切关联起来。这一转变的早期著名典型要数罗斯·贝尔坦了，在18世纪晚期她为路易十六王后玛丽·安托瓦内特以及众多的欧洲、俄国贵族创作服装与配饰。她被称为"款式商人"（*marchandes des modes*），意为她给礼服加上不同的装饰。不过，"款式商人"这一角色开始转变，部分原因在于贝尔坦在打造时尚造型方面拥有强大的实力。她从当时的重大事件中汲取灵感，比如，她精心制作了一件融合了热气球元素的头饰，以此向孟高尔费兄弟1780年代的热气球飞行致敬。她靠这些金点子为自己带来知名度，虽然同期的埃洛弗夫人与穆亚尔德夫人这些"款式商人"名气也很大，但当时巴黎时装鼎盛之势的最佳诠释者始终是贝尔坦。

1776年，法国以新型公司体制替代了原来的行会制度，提升了"款式商人"的地位，允许她们制造服饰，而不再只是装饰点缀。贝尔坦成了这一公司的首位"大师"，这大大增强了她在时装领域的声望。她为"大潘多拉"制衣，给这具人偶穿上最新款的时装，然后送往欧洲各地市镇和美洲殖民地进行展示。这是在定期出版的时装杂志出现之前进行时装宣传的一种主要方式。通过这种方式，贝尔坦助推了巴黎时装的传播，确定了它在女装界的主导地位。她发展出的广泛客户群以及她跟法兰西王后间的密切关系都确保了她在时装界的地位。有一点值得注意，当时的评论家们惊恐地发现，从贝尔坦的言行举止来看，好像她与她的贵族客户地位相等一般。她的阶层跃升也完成了一次重要的变革，为更多设计师们掌握话语权提供了平台。她很清楚自己的影响力，也笃信自己作品的重要性，她在创造时尚，也在为自己的顾客们打造时尚形象，后者将自己的时尚引领者地位托付给了贝尔坦。的确，她位于巴黎的精品店"大人物"（the Grand Mogul）大获成功，然后又在伦敦开了分店。她革新的风格搭配以及对于历史与时代热点的绝妙借用充分显示了她的设计实力，同时也说明她对于打造宣传攻势的重要性早已了然于胸。她因而成为高级时装设计师的先驱，她们将在19世纪的时尚主导群体中获得一席之地。

　　法国大革命短暂中断了巴黎时装的相关资讯往全球各地的传播。不过，革命甫一结束，法国的奢侈品贸易便迅速重建起来，而各家制衣师都开始力证自己的服装才最为时髦。路易·伊波利特·勒罗伊一手打造了约瑟芬皇后和其他拿破仑宫廷女性，以及一大批欧洲贵族的时装风格。1830年代，维多琳等人开始声名鹊起，其地位远在那些默默无闻的制衣师阶层之上。勒罗伊和维多琳，一如在她们之前的贝尔坦，都致力于创造新款设计，开创新的时装风潮，并不断巩固自己的杰出地位，同时帮她们那些有头有脸的老顾客维持显赫声名。但是，绝大多数制衣师，甚至是那些不乏贵族顾客的人，都没能实现设计原创。实际上，她们只对既有款式进行重新组合排列，以适应不同的顾客穿着。各种款式则从最知名的制衣字号那里或者从时装图样上抄袭而来。

　　不过，除了出现行业领先的女装制衣师之外，时装产业还有另外一面对时装设计师这一概念的演变产生了影响。艺术史学家弗朗索瓦丝·泰尔塔·维蒂有过记述，一些艺术家的工作方式类似于今天的自由职业设计师，因为制衣师们会向这些艺术家购买非常详细的时装图样。这些图样将会作为服饰的模板被使用，甚至可能被当作样品直接送到顾客手中。各家制衣师的广告也会附在这些插画的背面，同时标出画中服装的价位。到了19世纪中叶，查尔斯·皮拉特等艺术家开始以"时装与服饰设计师"宣传自己，刊载在当时的巴黎黄页中的"工业设计师"目录中。

整个西方世界也开始形成这个观念，认为服装需要由时装的权威人士来设计，运用特定的技法来明确廓形、剪裁以及装饰。每个市镇都可能有自己最时尚的制衣师，当时装开始随着大众对新款式的渴望而越来越快地变化时，时装设计本身的商业价值也在同步增加。时装设计师这一概念的最终明确不单需要随时可以推出新时装的富有创造力的个人，还要依靠人们对新鲜感与革新不断增长的需求。19世纪见证了中产阶级和富有实业家们的崛起，他们刚刚建立的地位在一定程度上是通过视觉展示来构建的，不管是他们的居所，还是更为重要的他们身上的服饰。高级定制时装开始成为更大的女性群体获得专属与奢华体验的一种渠道，其中美国人在19世纪下半叶成为这个群体中最大的客户群。

在这些变化之外，还有时尚媒体、摄影的发展，以及19世纪末出现的电影，它们使时尚形象得到了空前的广泛传播，且激发了女性对更为丰富多样及更迭迅速的时装款式的欲望。当城市的巨大发展带来更多的个性泯灭，时装成为能够构建身份，并让社会、文化及经济地位一目了然的重要途径。它同时也是愉悦与感官体验的来源，而巴黎的高级定制时装就是这一幻梦与奢靡王国的顶峰。

当"工业设计师"为规模较大的女装制作行业提供着时装设计的时候，高级时装设计师的演化最终构建起时装设计师的定位与形象。尽管1850年代最为著名的高级时装设计师查尔斯·弗雷德里克·沃思的成功一方面是依靠他扎实的生意经营，但在其商业努力之外人们看到的更多是他在创新设计中的精彩表达，还有他作为创意艺术家的权威身份，人们选择无条件地追随他的时尚宣言。作为一个在百货公司女装制衣部打磨技艺出身的英国男人，他在自己的职业生涯之初得以出众，一定程度上归因于他是一个从事由女性主导的职业的男性。其实，在1863年2月的《全年》(*All the Year Round*)杂志里，查尔斯·狄更斯就表达过对"嘴上带毛的女帽师"这一现象的厌恶。身为男人，沃思能够以不适用于女性的方式来宣传自己，他也能另辟蹊径地接待他的女性顾客，不考虑她们的身份阶层。他最知名的设计中加入了象牙色薄纱制作的泡泡纱，让人穿起来犹如云绕肩头，各层之间的珠饰与亮片在烛火辉煌的宴会厅中折射出熠熠动人的光辉。

其他高级时装设计师的名望也在不断增强，通常还因他们的皇家客户而日渐声名显著。在英国，约翰·雷德芬针对这一时期女性的角色转变，制作了以男士西装为设计基础的高级定制礼服，还为帆船运动制作了运动套装。在法国，让娜·帕坎等女性高级时装设计师制作出能够修饰女性身体的服装，完美体现了理想中巴黎女人的形象。许多顾客来自美国，因为巴黎依旧引领着时尚。一方面为了提升设计师的地位，一方面为了提

图3 保罗·普瓦雷精致的帝政风高腰线礼服裙,乔治·勒帕普1911年绘

供一种具有辨识度的身份与个性以推介自己的品牌,各家时装店明确了高级时装设计师即是革新者与艺术家这一观念。塞西尔·比顿形容英王爱德华时代的女性在努力不使自己的制衣师名字泄露。这些女性希望因为自己的时尚感而受到肯定,并始终保持自己的知名度高于其时装师。不过,高级时装店已经发展出它们独有的招牌款式,这些服装赋予了穿着它们的女性以鲜明的时尚地位。

20世纪的前几十年里,保罗·普瓦雷和露西尔等设计师开始蜚声国际。他们为戏剧明星、贵族及富豪们制作服装,并且宣扬着他们自身作为颓废的社会名流的身份。普瓦雷此时已经是现代意义上的时装设计师。他以标志性的奢华风格以及他创造的那些廓形逐季变化且造型夸张的款式而闻名于世。乔治·勒帕普笔下的时装插画生动展示了普瓦雷那条1911年的著名帝政风高腰线礼服的廓形,这件设计作品背弃了英王爱德华时期紧身束腰的时装样式。他那些布满刺绣的晚礼服与歌剧院外套汲取了从现代主义到"俄罗斯芭蕾"①等各种当代艺术与设计灵感,那些强有力的高级时装形象的格调又通过他自己的香水产品销售传播得更为深远。普瓦雷同时代的从业者们同样熟稔于利用现代广告与市场营销手段来构造他们各自时装店的形象。大多数人都将自己的设计销售给美国批发商,由他们就买下的每一个款式裁制出数量严格限定的服装。除了为单人定制的各款服装的销售所得,这类销售也给时装店带来了收益,而前者才是高级时装的本义所在。

---

① "俄罗斯芭蕾"(Ballets Russes)是巴黎的一家芭蕾舞巡演公司,于1909年与1929年间在整个欧洲巡演。——编注

两次世界大战之间是高级时装的一次鼎盛期,这一时期的玛德琳·维奥内、艾尔莎·夏帕瑞丽、可可·香奈儿等人通过她们的创作定义了现代女性这一概念。她们的成就强调了这样一个事实:长期以来,时装一直都是为数不多的女性能够以创新者与实干家身份获得成功的领域之一,她们领导着自己的业务,同时为无数女性提供自己高级时装工作室的工作机会。确实,高级时装是一项集体协作的事业,大牌时装店由许许多多的工作室构成,每个工作室负责一款设计的不同内容,比如说裁片、立体剪裁或串珠与羽毛等各类装饰。尽管每一件服装都有很多人参与创作,设计师们的想法始终是与作为独立创意个体的艺术家的理念保持一致的。这在一定程度上是因为设计与革新是时装中最受重视的两个方面,毕竟它们是各个系列的基础,又被视为整个过程中最具创新性的要素。特别是,这种对于个体的强调同时也成了一件有效的宣传工具,因为它强调了一个时装品牌的身份,而且真正为时装店提供了一张"脸面"。

　　尽管不受巴黎高级时装业中施行的那些严格条例的管理,其他国家也发展出了各自的高级时装设计师与定制产业。例如1930年代的伦敦,诺曼·哈特内尔和维克多·斯蒂贝尔就强调自己是时装设计师而不仅仅是王室制衣师。在纽约,华伦蒂娜(Valentina)逐渐发展出一种极其简洁的风格,通过吸收现代舞元素来创造一种美式时装身份。而在1960年代的罗马,华伦天奴(Valentino)倡导了一种与众不同的意大利式高级时装,崇尚极度女性化的奢华风格。

　　进入战后时期,纺织品与劳动力成本上涨使得高级时装更加昂贵。克里斯汀·迪奥等设计师在经历了1940年代的艰难之后，开始重新沉迷于繁复铺张，他们重视传统的高级定制工艺，在之后的十年里引领着高级时装对全球时装潮流始终如一的统治。从1960年代开始，尽管抛弃型的青少年时装开始兴起，成衣设计师的全球声望也在增长，但高级时装仍然停留在大众的视野里。其重要性虽然有所变化，但迪奥的约翰·加利亚诺、浪凡的阿尔伯·艾尔巴茨、香奈儿的拉格斐等几个特别的设计师依旧能够创造出在整个市场所有层面广泛传播的时装。尽管客户数量在不断下降，但成衣产品线、配饰、香水以及为数众多的授权产品仍将高级时装推到了巨大的全球奢侈品市场的头排位置。虽然欧洲的高级时装消费者变少了，但其他市场一直处在此起彼伏的兴盛之中。石油财富让1980年代的中东高级时装销量大幅增加，美元正值强势且人人钟爱显摆的里根时代的美国也是这样，而后共产主义时代俄罗斯创造的巨额财富则贡献了21世纪之初更多的客源。再加上名流文化的显著与红毯礼服的兴起，高级时装设计师们继续制作着每一季的服装系列。即便这些孤品式的设计本身并不能创造什么利润，但它们带来的大量关注巩固着处于高级时装产业核心位置的设计师们始终如一的重要性。

成衣设计师的演化

英国时装记者艾莉森·赛特尔在她1937年的《服装线》一书中写道，巴黎高级时装行业内部相互联系的性质是行业兴盛的关键。面料、服装以及配饰的设计师和制作商们相互间均密切联系，可以对各自领域内的发展做出及时反应。

流行趋势因此得到迅速鉴定，然后收入高级时装设计师们的服装系列之中,这使得巴黎始终保持着它在时装业界的领先地位。令赛特尔印象深刻的还有时装在法国文化中的深植,所有社会阶层的人们都对穿衣打扮和时尚风格抱有兴趣。就像赛特尔所写,高级时装设计师们"通过观察生活来预测时尚",而这一方法在成衣设计师的演化过程中显得尤为重要。高级时装设计师们发现许多女性希望购买的不仅是紧随当下风格潮流的衣服,这些服饰还得是由时尚品牌设计生产的。

自1930年代初期起,设计师们开始陆续推出一些价位略低的时装系列,以触达更多的受众。吕西安·勒隆就是一个例子,他开始制作自己名为"勒隆系列"的产品线,销售的成衣价格只有他高级时装系列定价的零头。高级时装设计师们一直开发着成衣服饰,比如1950年代雅克·法特就曾为美国制造商约瑟夫·哈尔珀特设计过一个极为成功的产品系列。不过,当皮尔·卡丹1959年在巴黎春天百货发布他的成衣系列的时候,他却遭到了巴黎高级时装公会的短暂除名,这个协会的存在本身是为了规范高级时装产业,而卡丹被除名的原因就是他没有获得其许可便这样设立了自己的分支产品线。与此同时,卡丹也在积极探索远东地区的潜在市场,谋求在全球范围内实现商业成功。与他大胆、现代的设计风格联系起来观察,他的这些举动实际上都是法国时装业重心转移的一部分,高级时装设计师们竭力保持着他们的影响力以应对越来越多斩获成功的成衣设计师。1966年,伊夫·圣罗兰的"塞纳左岸"(Rive Gauche)精品店开业,以长裤套装与色彩灵动的分体款时装来响应流行文化并认同女性社会角色的转变。圣罗兰证明了高级时装设计师同样可以通过其成衣系列引领时尚。在1994年艾莉森·罗斯索恩对一位名叫苏珊·特雷恩的顾客的采访中,这位女士形容圣罗兰新的时装线"令人十分激动。你能买下整个衣橱:任何你需要的东西都买得到"。然而,1960年代一般被视为大规模生产、年轻化的成衣开始以前所未有的姿态引领时尚的关键节点。美国的邦妮·卡欣等设计师,英国的玛丽·奎恩特等人,还有意大利包括璞琪在内的设计师们当时都在市场的各个层次里确立着自己的时尚影响力,塑造了时装设计、销售以及穿着的新范式。

虽然成衣服饰自17世纪开始就脱离巴黎高级时装独立发展,但直到1920年代它们才开始真正因为自己的时尚价值而不是价位或质量被设计销售。在巴黎城中,这意味着高级时装设计师们需要在未来的数十年里与全球的百货公司达成协议,让他们销售自家专属版本的高级时装,同时为其开发成衣产品线。而在美国,包括汤利制衣(Townley)在内的生产商以及萨克斯第五大道精品百货等百货公司迅速雇用设计师为自己匿名研发时装产品线。

到了1930年代,这些设计师开始走出无名的幕后世界,将自己的名字纳入品牌信息。在纽约城里,精品百货罗德泰勒副总裁多萝西·谢弗,推行了一系列将店中美式成衣与定制时装设计师两者并列推广的宣传活动。橱窗与店内陈设中加入了有名有姓的设计师的照片,并与其设计的时装系列一同展示,鼓动一种曾经只有高级时装设计师们才能享受的个人崇拜。这也是人们努力培育本土成衣设计人才实践的一部分,毕竟大萧条带来的困境使前往巴黎获取时尚资源的洲际旅行显得太不经济。同时它也充分说明时装设计师需要依靠抱团组队来提升其自身时尚资质的行业认可度。巴黎继续保持着它作为时尚中心的地位，但到了1940年代，法国的行业影响在战争里中断了，纽约开始确立起自己的时尚地位。再往后发展，全球的众多城市循着同样的发展过程，投资自己的时装设计教育，举办各自的时装周以推广本土设计师的时装系列，并力图销往国内国际两个市场。

21　时装设计师在这一过程中的角色至关重要，他们还提供着创造的推动力，外加颇具辨识度的面孔，后者可以作为推介宣传的坚实基础。1980年代，安特卫普与东京展示了各自培养出众设计师的实力，一批设计师开始成名，其中就有比利时的安·迪穆拉米斯特和德赖斯·范诺顿，以及日本籍的Comme des Garçons[①]品牌设计师川久保玲和山本耀司。到了21世纪初期，中国和印度也像其他国家一样，开始投资自己的时装产业，并培养自己的季节性时装秀。

　　设计师们所接受的训练方式直接影响着他们创作时装系列的方法。例如，英国艺术院校强调研究和个人创意的重要性。这种对创造过程中艺术元素的重视造就了像亚历山大·麦昆这样的设计师，他们从历史、典雅艺术与电影中汲取灵感。麦昆的设计系列通过主题鲜明的场景呈现出来，模特有时在一个巨大的玻璃箱子里扭动着身体，有时又要一边随着旋转平台缓缓转动一边任由一只机械喷嘴喷绘。他的模特们被打造为人物角色，有些像通过服装和布景娓娓道来的叙事。麦昆那种电影式的展示方式在他2008年[②]春季时装系列中一览无余，这一季他以1968年的老电影《孤注一掷》（*They Shoot Horses Don't*

---

① 后文统一简称CDG。——编注
② 实为2004年，应为讹误。——译注

They? ）为灵感来源。这场秀营造出一种大萧条时代马拉松式舞会般的场景主题，请来了先锋舞蹈家迈克尔·克拉克进行编舞。模特们轻盈地穿过舞池，身着灵动的茶会礼服和旧工装裤，她们的肌肤汗光闪闪，双眼神色略散，似乎已在男舞者的半举半曳之下跳了一个又一个小时。麦昆以一种空前震撼的场面来宣传自己的时装，这进一步巩固了他同名品牌的成功，同时也充分证明了他那些无限创意的魅力。

与此相反，美国的院校则喜欢鼓励设计师专注为特定的顾客人群创作服装，并将商业考量以及制造的便捷性始终放在第一位。他们以工业设计为模型以强化一种民主设计的理念，目标指向最大数量的潜在消费者。从1930年代到1980年代，邦妮·卡欣等设计师的作品很好地例证了这种方法如何能够产生标准化的时装系列以针对性地满足女性的穿衣需求。她的设计看上去线条流畅，同时显示出对细节的密切关注，并搭配上有趣的纽扣或腰带搭扣让它们简单的廓形活泼起来。1956年，卡欣曾告诉作家贝丽尔·威廉姆斯，她坚信一个女人衣橱的75%都是由"永恒经典"单品组成的，她还特别说明"我的全部服装都是极其简单的……它们就是我想穿在自己身上的那种衣服"。她为工作、社交还有休闲等各种场合设计生活方式服饰，同时把自己宣传为她的方便穿着（easy-to-wear）风格的具体实例。这种类型的设计开始逐步塑造美式时尚的个性特征，不过它这种简单的特质也可能使得一个品牌难以确立自己独特的形象。从1970年代末到1990年代末，卡尔文·克莱恩运用极具争议的广告为他的服装与香水系列进行宣传。典型的形象就有1992年的"激情"（Obsession）香水广告中裸露且雌雄莫辨的少年凯特·莫斯,这些画面营造出一种前卫、现代的品牌形象,这实际上与他在许多设计中的保守风格大相径庭。

这些设计师一直坚持个人担纲时尚原创者的理念,而与此同时,许多时装店开始雇用完整的设计师团队为其制作产品系列。正是出于这一原因,比利时设计师马丁·马吉拉一直拒绝接受个人采访,同时竭力避免出现在镜头里。所有报道与媒体通稿中一概署名"马丁·马吉拉时装屋"(Maison Martin Margiela)。2001年,在时装记者苏珊娜·弗兰克尔对马吉拉时装屋另类的时尚操作方式进行的一次传真访问中,启用非专业模特这一做法被解释为品牌全盘策略的一部分:"我们绝对不是要反对作为个体的职业模特或'超模',我们只是觉得更希望将关注点聚焦在服装上,而媒体和他们的报道并没有把全部关注给予它们。"他的品牌标签一般留白或者只是打上该款服装所属系列的代码。这种做法转移了人们对于个体设计师的关注,同时暗示了要完成一个时装系列必须要靠众人通力合作,同时这种做法也使他的作品与众不同。对于另一些设计师来说,他们的重心更多地放在名流顾客群上,这些人给他们的时装系列罩上了一层靓丽迷人的光环。21世纪初,美国设计师扎克·珀森就受益于包括娜塔莉·波特曼在内的年轻一代好莱坞明星们,她们身穿他的礼服走上红毯。在这类活动中,明星们获得的媒体报道足以带动新晋设计师的作品销售,同时创建他们的时装招牌,就像朱莉安·摩尔穿上圣罗兰旗下斯特凡诺·皮拉蒂的设计后成功地将他推上了赢家的位子。

男装设计师们也开始在20世纪不断崛起至行业前沿,尽管他们一直没能引发与女装设计师们相同水平的关注度。男装设计一般集中在西服套装或休闲服饰上,且人们普遍认为男装缺乏女装可以被赋予的那种宏大壮观而又激动人心的东西。不过,男装设计师的代表人物还是从1960年代开始逐步出现了,比如伦敦的费什先生与意大利的尼诺·切瑞蒂。两者都充分发掘了那个年代明艳的设计风格,在他们的时装设计中运用了生动的色彩、图案和中性化的元素。在1966年开出自己的精品店之前,迈克尔·费什在萨维尔街精英式的工作环境中发展出了他自己的风格。与此同时,切瑞蒂则自他家族的纺织品生意中培养出了自己线条明快的设计风格,最终在1967年推出了他个人的第一个完整男装系列。巴黎高级时装设计师们同样开拓出男装设计,包括1974年的伊夫·圣罗兰男装。1980年代,设计师们继续探索着男装设计的各种规范,将精力集中到对传统西服的改良上。乔治·阿玛尼剥除了传统男装硬挺的内衬,创造出柔软无支撑的羊毛和亚麻西装外套,而薇薇恩·韦斯特伍德则挑战了时装性别边界的极限,给西装外套缀上亮珠与刺绣,或者让男模们穿上短裙和裤袜。

从1990年代开始,德赖斯·范诺顿时装系列中丰富的色彩与质感,还有普拉达(Prada)设计中革新性的纺织面料,都充分展示了男装设计能够以微妙的细节引人注目。男士美容与健身文化的发展也给这一领域带来新的关注。21世纪之初,拉夫·西蒙等一批设计师,尤其是在2000年到2007年之间为迪奥·桀傲男装(Dior Homme)工作的艾迪·斯里曼,为男性开发出一种纤瘦的廓形,影响极为深远。斯里曼极窄的裤型、单一的色调、极为修身的西装外套意味着它们必须穿在透露出中性气质与特立独行态度的年轻身体上。各界名流、摇滚明星还有

图4 艾迪·斯里曼2005年春季系列中极具影响力的紧身廓形

高街品牌店铺，迅速地接受了这一造型，充分展示了自信的男装设计也可以拥有强大的力量与影响力。

亚文化风格一直是1960年代以来男装系列中最为重要的一类借鉴元素。从六十年代"摩登族"（Mods）穿着的紧身西装到八十年代休闲派身上色彩柔和的休闲装，街头风格平衡了个性与群体身份。它因而吸引许多男性开始寻求既能起到统一着装作用，又能让他们自己加以个性化改良的服装。各种亚文化群体的成员通过他们的穿着风格以多样的方式塑造着自己，有的靠定制专属服装来完成，有的则靠打破那些规定应当如何穿着搭配的主流规则来实现。到了1970年代晚期，这种自己动手的风气在朋克族（Punks）身上体现得最为明显，他们在衣服上添加标语，别上别针，撕开衣料，创造出自己对经典的机车皮夹克和T恤衫的演绎。1990年代中期开始，日本的少男少女们也开始制作自己的服装，给它们加上和服腰带等传统服装元素，创造出多变的服装风格，而这些服装又都符合他们对夸张与梦幻的热爱。通过借鉴这些做法，时装设计师们得以为自己的设计系列注入一种看起来叛逆感十足的前卫感。

的确，从1990年代开始，时装消费者们就越来越追求通过定制服装与混搭设计师单品、高街款式与古着衣饰来实现自己造型的个性化。这让他们当起了自己的设计师，哪怕没有那么多件可改的衣服，也要把整体造型和形象打造成自己想要表现的样子。1980年代的"时尚受害者"这一概念，即从头到脚全穿同一设计师的作品的人，促使许多人采取行动，力求通过在自己身上进行改良与造型搭配来表现他们的独立创意，而不是依赖设计师们为他们构建起某种形象。这种做法效仿了亚文化风格

图5 日本的街头时尚融会借鉴了东方及西方、复古与新潮的丰富元素

以及专业造型师的工作模式。它反映出某些消费者群体中不断成熟的自我认知,还有他们心中既想成为时尚潮流的一分子又不甘任其摆布的愿望。20世纪无疑见证了大牌设计师逐渐成长为时尚潮流的引领力量,但他们也不断地迎来新的挑战。1980年代或许是设计师个人崇拜的顶峰,即便许多品牌如今依旧备受崇敬,它们现在却必须与数量空前的全球对手同台竞技,同时还要对抗众多消费者想要设计个人风格而不是一味遵从时尚潮流的强烈愿望。

28

第二章

艺　术

　　安迪·沃霍尔1981年的作品《钻石粉鞋》(*Diamond Dust Shoes*)呈现了在漆黑背景下杂乱摆放的色彩明亮而鲜艳的女士浅口便鞋的画面。作品以照片丝网印刷为基础,鞋子从顶角拍摄,观者犹如正在俯视衣橱隔板上零散的一堆鞋子。一只炫目的橘色细高跟鞋耸立在一只端庄的番茄红色圆头鞋旁边,而另一只深蓝色缎面晚装鞋旁边是一只橙红色带蝴蝶结的船形高跟鞋。所有的色彩都是逐层叠加在画面之上的,制造出一种将众多风格与造型的鞋子拼凑在一起的卡通效果。

　　照片的剪裁给人营造出一种这堆鞋子难以计数的印象,例如一只淡紫色靴子只露出一点鞋尖可见,从画框边缘伸进了画面。这一图像经过了艺术家精心的组织;即便看上去杂乱,但每一只鞋都是巧妙陈设的,并露出数量刚好的鞋内商标可见,以此强化它们的高端时尚地位。画作令人想起时尚大片与鞋履店铺,并以此指涉了对时尚来说基本的视觉与真实消费行为的统一。沃霍尔的画作在丙烯颜料的光泽下显得平整光洁,这种光

图 6 安迪·沃霍尔 1981 年的《钻石粉鞋》展现了充满魅力与诱惑的鞋履设计

泽感在整幅画面上撒开的"钻石粉"的作用下得到了加强,这些"钻石粉"折射光线时让观者眼前一片熠熠生辉,炫目迷人。作品闪闪发光的表面明确地展现了时尚光鲜亮丽的外表以及它能够改变日常生活的能力。

1950年代末，沃霍尔以商业广告艺术家的身份工作，他的客户就包括I.米勒制鞋。他为其绘制了众多妖娆、明快的鞋履作品，尽显魅惑之态。他自身与商业的联姻还有对流行文化的热爱意味着时尚对他来说是一个理想的主题。时尚在沃霍尔的丝网印刷画和其他各种作品中占据着重要地位，而他也不断运用着装与配饰，包括他那著名的银色假发来改变自己的身份。1960年代，他还开了一家精品店，起名"随身用品"（Paraphernalia），混搭销售一众时装品牌，比如贝齐·约翰逊（Betsey Johnson），还有福阿莱与图芬（Foale and Tuffin）。"随身用品"精品店的开幕仪式上还请来"地下丝绒"乐队演出，因而将沃霍尔在不同方向上的开创性艺术创作综合在了一起。他深刻理解到在这十年里，时尚、艺术、音乐与流行文化之间已经结盟。将先锋流行音乐与基于色彩明艳的金属、塑料以及撞色印花的抛弃型和实验性服装进行融合，不仅表达出这一时代的创造激情，还帮助确立了时尚标准。对于沃霍尔来说，艺术或设计形式不分高低。时尚从不因其商业上的紧迫性或者说它的短暂易逝而受人唾弃。相反，这些固有特性在他的作品中被大肆宣扬，作为他对当代快节奏生活的着迷的一部分。于是，《钻石粉鞋》令人炫目的表面赞美了时尚对于外在美好与华丽的关注，而他的精品店则吸引人们关注处于时装业最核心的商业行为与消费主义驱动力，很多当代艺术市场的核心也确实是如此。在沃霍尔的艺术作品里，时尚固有的短暂易逝而又物质至上的缺陷，变成了对孕育时尚之文化的评述。对沃霍尔来说，大众文化与高端奢侈品的各种元素可以和谐共存，正如它们在各类时尚杂志或各种好莱坞电影中共存一样。在他的作品里,复制品与孤品拥有相同的地位,他也能非常轻松地在不同的媒介间转换,对电影和丝网印刷或平面设计的各种可能性怀有同样的热情。沃霍尔既不认为这些限制了他的创作,也不觉得应将商业剥离艺术以保持它的合理性,而是拥抱了各种矛盾体。在他1977年的《安迪·沃霍尔的哲学:波普启示录》[*The Philosophy of Andy Warhol (From A to B and Back Again)*]一书中,他写到推动着他的艺术前进的种种交融难厘的界限:

> 商业艺术(business art)其实是紧随艺术之后的一步。我开始是一个商业广告艺术家(commercial artist),而现在我想最终成为一个商业艺术家(business artist)。当我从事过"艺术"——或无所谓叫什么——的那件事后,我就跨入了商业艺术。我想成为一个艺术商人,或是一名商业艺术家。商业上的成功是最令人着迷的一门艺术。

自19世纪中叶起,时尚就加快了步伐,触及更多的受众,拥抱工业化的生产过程,并运用多种引人注目的手段完成商品销售。艺术经历了同样的变革进程:艺术市场不断成长,开始拥抱中产阶级,机械化复制改变了艺术只能专属于某个藏家的观念,艺术机构及私人画廊重新思考了艺术品陈列与销售的方式。时尚和艺术的主题之间同样存在着相互交叉,从身份与道德两大议题,到对艺术家或设计师在广泛文化下被感知方式的关心,再到对身体的表现与运用的关注。

有时时尚也会被呈现为艺术,但这也引出一连串问题。一些设计师在自己的作品中化用了艺术创作方式,但他们仍身处时装产业框架之内,同时运用这些借鉴来的方法探索时装的本质。例如,职业生涯早期的维果罗夫就立志专注举办时装秀,而不愿意制作可供销售的衣服,他们的设计于是都成了孤品或少量制作的款式,它们的存在只是为了佐证那场时装秀在时装体系中的重要性,而非成为实穿的服装。尽管如此,他们的作品依然占据着时装界的一席之地,接受着时装记者们的热烈评议。这似乎是在为他们后面推出的几个时装系列展开宣传活动,这些后续的设计系列被投入生产。他们的作品同样强调了不同类型设计师之间的差异。维果罗夫对时装的诠释融入了他们对时装秀角色的迷恋,以及它检验景观与展示边界的潜力。在自己这些时装系列的秀场呈现中,他们游走于艺术、戏剧以及电影之间。2000年秋冬系列发布时,两位设计师慢慢地给一位模特套上一层又一层的衣服,最终为她穿上了整个系列的服装。这其实呈现了在真人身体上完成试衣的过程,它是传统时装设计最核心的部分。最后包裹着女模特的那些尺寸夸张的时装成品看上去几乎将她变成了一个僵硬的人偶、一个活着的人体模型,以及设计师手中的玩物。在2002/2003年的时装秀上,所有服装都是明艳的钴蓝色,在秀中充作电视电影特效拍摄中使用的蓝幕。他们将电影投影在模特的身体上,模特们的身形尽失,在电影画面投射于身体表面时他们的身体似乎也在明灭闪烁。

在维果罗夫的设计与秀场展示中,艺术化方式的运用为时装实践增色良多,但这未必能把他们的时装转变为艺术。他们的作品在国际时装周的大环境中展示,指向的是时尚群体,强调的是服饰与身体交流的方式。哪怕在还未将自己的服饰投入批量生产的阶段，他们也是遵循着时装季设计发布，还有尤为重要的一点，他们一直忠于时装的基础元素：面料与身体。

有时人们将时装与艺术相提并论，以赋予它更高的可信度、深度以及意图。不过，比起对时装真实意义的揭示，这种做法也许更多地暴露了西方世界对于时装缺乏这些特质的担忧。当一条1950年代的巴黎世家（Balenciaga）连衣裙陈列在画廊的古旧玻璃展柜中，它也许看起来会像一件艺术品。不过，为了展现它的价值或创造它时所运用的技艺，它并不需要被称为一件艺术品。如同建筑等其他设计形式，时装也有着自己独特的追求，如此便避免自己成为纯粹的艺术、技巧工艺或是工业设计。它实际上更像是一种吸纳了所有这些方式的各种元素的三维设计。巴伦西亚加先生严苛要求精确的形态，为面料的褶裥与结构带来平衡感与戏剧化效果，再辅以工坊里工人们的精湛手工，让这一袭裙子变成了一件卓越的时装。它并不需要被冠以艺术之名来佐证其地位，艺术这一字眼忽略了在渴望创造并挑战时装设计界限之外，巴伦西亚加制作裙子的原因：希望装扮女性，并最终卖出更多设计。我们不应认为这一层诉求削弱了他所取得的成就，反倒应该看到，它帮助我们进一步理解了巴伦西亚加的创作方式，他全力开拓种种"边界"，以创造能够给予穿衣人与观衣人同等启迪的时装。

理解时装应当以它自己的语境为基础，这使得时装与艺术、文化其他方面的相互交织更加妙趣横生。它为艺术、设计以及商业在一些时装实践者的作品中发生联系、交织重叠开辟了新的路径。的确，时装能够如此令人着迷而又令某些人感觉难以捉摸的一个重要原因就是，时装总是能侵吞、重组并挑战种种既有定义的边界。因此,时装能够突出关于一种文化中何为可贵事物的矛盾。安迪·沃霍尔和维克多与罗尔夫等各种各样的设计师与艺术家们都在利用文化矛盾和态度创作作品。对时装来说,关注身体与面料,以及它常常以穿着与销售为设计目的这一事实,将其与纯艺术区别开来。不过,这并不妨碍时装拥有自己的意义,而艺术世界对时装一如既往的迷恋正突显了它在文化上的重要性。

肖像画与身份塑造

时装与艺术之间最显而易见的联系大概就是服装在肖像画中扮演的角色。16世纪,宗教改革在北欧的影响致使宗教绘画的委托需求下降,艺术家们因而转向新的绘画主题。从文艺复兴开始,人文主义对个体的关注让众多的贵族成员开始希望由艺术家为自己绘制肖像。肖像画的成长建立起艺术家与画主、时装与身份表达之间的关联。霍尔拜因笔下的北欧皇室与贵族成员肖像画探索了油画所能够传达的视觉效果,表现出绸缎、丝绒以及羊毛等不同面料之间的质感差异。霍尔拜因精细的绘画水平直接体现在他为肖像画准备的详尽底稿上。珠宝首饰的精巧细节在他的素描里被悉数摹画,构成女士头饰的精致的层层平纹棉布、亚麻布以及硬质衬底在他的画笔下与主人公的脸庞和表情一样细腻。霍尔拜因深深明白时髦的衣饰对于表现他这些主顾们的富有与权势以及性别与地位发挥着重要的作用。这些属性在他的画作中被清晰地呈现出来,不仅成为过去服饰风格的纪念,又令人想起时装在构建一种能让同时代人轻松领会的身份时曾经扮演的角色。他所绘的亨利八世系列肖像展现了当时追求的视觉上的宏大，利用层层垫高的丝绸与锦缎来增加身形大小与威严之感。黄金与珠宝的镶边与配饰进一步增强了这种效果，外层的面料适度裁剪收短，目的就是要露出里面更加华贵的服饰。他绘制的女性肖像同样细节丰满。甚至在他1538年为丹麦的克里斯蒂娜绘制的风格沉郁的肖像画中，主人公所穿丧服面料的华美依旧一览无余。光线洒落在裙面上深深的褶皱和宽松堆叠的两肩上，强化了她这身黑色缎面长裙所散发的柔和光泽。与之鲜明对比的，是由领口直贯下摆的两道棕红中略泛茶色的皮草，还有她手握的一双软质浅色皮手套。与这一时期欧洲各地的其他艺术家一样，霍尔拜因的画面构图将焦点集中在主人公的面部，同时又极为重视对他们华丽服饰的展示。

从意大利的提香到英国的希利亚德，面料与珠宝的华美在各个艺术家的作品中呈现。甚至当服装素净又未施装饰时，如在丹麦的克里斯蒂娜肖像中，材质的奢华在主人公身份地位的构建中也扮演了重要角色。这种展示的重要性对于当时的人们来说是非常容易理解的。那时的纺织品非常昂贵，因而备受珍视。有能力购置穿着大量的金线织物与丝绒面料充分显示出主人公的财富。层层外衣之下不经意露出的雪白的男士衬衣和女士罩衫，进一步强化了主人公的地位。清洁在当时也是身份地位的一种标记，不管是及时清洗亚麻布料保持其洁白，还是浆洗轮状皱领并且恰当地压出复杂形状，都需要大量的仆人。

艺术不仅仅为彰显皇室与贵族身份服务，它还表现着个性、审美，还有主人公与时装之间的关系。当霍尔拜因等艺术家奋力精确描摹当时的时装，并将之视为自己作品总体的现实主义手法的组成部分时，另一些艺术家则运用了更加华丽的创作手法。17世纪期间，凡·戴克等人笔下的主人公，常常身着各色垂褶的面料，它们以各种超乎自然的方式萦绕着他们的身躯。艺术家们用神话般的服饰塑造人物的身体，有意地让人联想起希腊众女神。女人们围裹着柔光绸缎，看上去如同飘飞在身体四周，浮动于肌肤表面之上。画中男人们身穿的服饰，一部分源于现实，一部分则是幻想虚构。虽然凡·戴克也画时髦的裙子，但他常常用自己一贯的审美对其进行改造，喜好反光的表面与连绵的大色块。这样一来，艺术便影响了时装，它不只记录了人们穿什么、怎么穿，更记录了人们理想中的美丽、奢华与品味。

当时装在18世纪开始逐季更迭，艺术与时装的关系变得更加复杂起来，有的艺术家开始担忧这种变化将会影响他们作品的重要性。一些肖像画师，例如约书亚·雷诺兹，追求作品经久不衰，希望创造出能超越自身时代的画作。时装似乎妨碍了他们实现抱负，因为它的存在能把一幅画拉回到其创作的时代。由于流行风格年复一年地改变，就算不是逐季变换，肖像画所属的年代完全可以精确判定。对凡·戴克和他画中的主人公而言，仿古典风格的服装一定程度上只是对梦幻华服的一种有趣尝试，对雷诺兹而言，它则是一种严肃的手法，试图实现与时装的割裂，并提出一种能确保肖像画对后世意义重大的另辟蹊径的方法。他因此竭力从自己的艺术中抹去时装的痕迹，让笔下的主人公裹上假想出的衣料，将人物形象与古代雕塑上可见的那种古典垂褶服饰融为一体。时装拥有一种强大的力量，它能够塑造人们对身体与美的认知，这被认为破坏了雷诺兹的意图。尽管他画出的裙子通常款式朴素，但18世纪最后25年里的时装大多如此，比如他所钟爱的拖地而修身的廓形。主人公对时髦形象的渴望同样阻碍了他崇尚古典的审美。女性画主们坚持戴上高耸并扑上白粉的假发，还经常点缀上羽毛装饰。她们的面庞同样搽成白色，双颊则施以时髦的粉红色脂粉。

主人公想让自己看起来时髦的诉求，再加上艺术家难以挣脱自己所处时代的主流视觉理念，这意味着想要绘制出一幅完全脱离自身时代背景的肖像画几乎是不可能的。安妮·霍兰德在她的《看透服饰》(Seeing through Clothes)一书中提出：

> 在现代文明的西方生活中，人们的着装形象在艺术中要比在现实生活中看起来更具说服力也更容易理解。正因为如此，服装引人注目的方式得到了当前着装人物画所做的视觉假定的调解。

霍兰德认为，通过艺术呈现能够"认识"到的不仅仅是身着服装的身体。她还认为艺术家们眼中的景象其实是同时代时装风尚培养起来的，甚至在描绘裸体时，身体的形态与展示方式都受到了流行时尚理念的调和。15世纪克拉纳赫笔下裸体上发育不足又位置靠上的双乳、位置较低的腹部，1630年代鲁本斯的《美惠三女神》(Three Graces)全身像，还有19世纪之初戈雅绘制的或着衣或赤裸的《玛哈》(Maja)，都证明了流行的廓形对人体形态描摹方式的深刻影响。在每个例子里，穿上服装之后的人体形态被塑身衣、衬垫以及罩袍重塑，又强加在画中的裸体人物体态上。所以，肖像画与时装的关系是根深蒂固的，也充分说明了视觉文化在任一时代都有着相互关联的本质。

这种相互的联系在19世纪变得更加显著，塞尚、德加还有莫奈等艺术家开始把时装图样作为他们笔下女性形象和她们所穿服饰的范本。由于许多人都是通过图像来了解时装的，不管是油画、素描、时装图样，还是后来的照片，观看者与艺术家一样，一直以这些图像为依据，被引导着理解他们在自己周遭所见的那些着装后的身体。实际上，艾琳·里贝罗进一步延伸这一观点，认为定制或购买艺术品这种行为中包含的物质主义一定程度上也是一种消费文化，后者见证了19世纪后半叶时装产业的成长，以及顶级肖像画家们与诸如查尔斯·弗雷德里克·沃思这样的高级时装设计师们价位相当的昂贵收费。里贝罗在她的《华服》（*Dress*，1878）一书中引用了玛格丽特·奥利芬特的话以佐证它们这种紧密的联系："现在出现了一个全新的阶层，她们的穿衣打扮紧随画像上的形象，当她们想买一件礼服时会问'这件能上画吗？'。"

时装与其图像呈现之间的模糊边界最引人注目的例子，大概要数皮埃尔-路易·皮尔森自1856年到1895年间为卡斯蒂廖内伯爵夫人维尔吉尼娅·维拉西斯拍摄的400余张照片组成的集子。她积极参与了自己的穿衣、造型以及姿势设计。她自己也因此充当着艺术家的角色，把控着自己在时装中的展示以及照片中的形象表达。她精心装点的19世纪中叶的各式裙子如同时装照片一般发挥作用，同时跨越时装图像的范畴建构起自己与衣服独特的关联。卡斯蒂廖内很清楚自己是在每张照片里进行表演，她将自己置于一个恰如其分的环境里，不管是在摄影棚里还是在阳台上。她充分示范了"自我造型"的魅力，运用服饰来定义或架构起人们对她的认知，以及她身体所展现出的形态。对她而言，时装与艺术之间的相互关联是一件强大的工具，让她

图7　19世纪中叶，卡斯蒂廖内伯爵夫人在大量摄影作品中创造出自己独特的个人形象

得以尝试丰富多样的身份,就像皮埃尔·阿普拉克西纳与格扎维埃·德马尔涅认为的:

> 卡斯蒂廖内对身体的运用——她的艺术创作的直接来源——以及她对自己公众形象的精心筹划组织[预示了]……诸如身体艺术与行为艺术等当代事物。

时装在视觉文化中扮演着意义非凡的角色,真实的服装与它们在艺术和杂志中的再现之间又存在着无法割裂的联系,这意味着艺术家们常常对时装所具力量的态度摇摆不定。温特哈尔特和约翰·辛格尔·萨金特等肖像画家利用主人公的时髦服饰塑造画面构图,同时也表明画主们的地位与个性,而另一类艺术家,其中以前拉斐尔派最为知名,则全力抵抗着时装对美、风格以及审美典范无孔不入的制约。到了1870年代,掀起了一场唯美服装运动,它力求为时装对身体的限定,尤其是塑身衣对女性身体的禁锢提供另一种选择。无论男女,都转而选择了已成为历史的版型宽松的时装风格。但是,唯美服装本身也成了一种时装潮流,虽然它明确了一个观点,即艺术家与纯艺术爱好者们应当穿得不一样些,应当拒绝时装流行风格。尽管他们可能会拒绝同时代的时装潮流,但他们对自己穿着打扮那种刻意的淡漠,反而含蓄地认可了时装能够塑造自己形象的功能以及服装塑造身份的力量。

跨界与再现

20世纪,艺术与时尚之间发生了大量的相互借鉴与跨界合作。高级时装不断演进的审美力将精湛的手工技艺与个别设计师的愿景和压力结合起来，打造了牢固的商业运作模式，以保障品牌的持久繁荣。高级时装设计师们努力地构建起自己的时装店在当代美理念下的个性特质，这必然会使他们将现代艺术视为一种视觉的启迪与灵感来源。在保罗·普瓦雷的实践中，这意味着一种对异域风情理念的探索，他像画家马蒂斯那样前往摩洛哥游历，寻找着与西方世界全然不同的对色彩与形式的运用。普瓦雷梦幻般的浓郁色彩块、垂褶的伊斯兰女眷式裤装，还有宽松的束腰外套共同搭建起一种女性气质的典范，而这种典范自19世纪晚期开始就越来越清晰地出现在大众流行与精英文化里。普瓦雷和妻子丹尼斯穿着东方风格的长袍、斜倚在沙发上的样子被摄影师记录了下来，照片背景就是他们所办的那次臭名昭著的"一千零一夜"派对。联系普瓦雷的设计一并审视，这些图像使他的高级时装店成为奢靡与颓废的代名词。重要的是，它们还将普瓦雷放在不妥协的摩登者的地位，尽管他的许多服装设计基础仍旧是古典元素。普瓦雷很清楚他需要利用独具创意的艺术家概念以培养一种个人形象，同时又制作出畅销国外尤其是美国市场的服装。与其他高级时装设计师一样，他的作品必须平衡好个人客户对定制服装的需求——这与纯艺术的独创性更为相似——以及创作适当款式销售给各国制造商进行复制的商业需要这两者之间的关系。虽然普瓦雷努力维系着一种艺术家的形象，也充分利用了"俄罗斯芭蕾"的影响力，但他还是踏上了前往捷克斯洛伐克和美国的宣传之旅，目的是要在更广泛的受众中提升自己作品的知名度。

南希·特洛伊也写到了20世纪前几十年里纯艺术创作与高

级时装之间的微妙关系。她发现，每一领域的变化都是对大众与精英文化之间越来越模糊的边界的反应，因而也是对"真正"原创与复制行为之间的差别做出的反馈。她认为，设计师和艺术家们试图"探究、掌控、引导（尽管未必一定能避开）商业与商品文化所谓的腐化影响"。

高级时装设计师们有各种各样的方式来恰当处理这些问题，并将当代艺术的影响力吸纳到他们的时装设计之中。在明艳撞色风格以及注重戏剧化个人呈现的影响下，普瓦雷的作品不断涌现。于是他与艺术家们进行直接的跨界合作也就不足为奇了，代表性的合作项目就有马蒂斯和杜飞的织物图案设计。首屈一指的先锋艺术家们与时装行业里的大牌设计师之间这样的联系看起来既理所应当又能互利共赢。任意一方都能够尝试、探究新方式去思考并呈现自己心中的理念。通过与另一种形态的前卫当代文化进行交融，以将视觉与物质相结合，各方都有可能从中受益。艾尔莎·夏帕瑞丽推出了更广泛的跨界合作，由她与萨尔瓦多·达利和让·科克托共同完成的设计最广为人知。这些联袂合作创造出的服饰赋予超现实主义信条以生命，其中就包括达利的"龙虾裙"。跨界合作把超现实主义艺术运动所热衷的并列手法，以及它与女性特质认知之间的复杂关系带进了有形的世界，穿着夏帕瑞丽设计的人们把自己的身体变成了关于艺术、文化以及性别的主张。

对玛德琳·维奥内而言，她对当代艺术理念的兴趣体现在她对三维立体剪裁服装技术的探索中，给她灵感的正是意大利未来主义碎片化的表达风格。她与欧内斯托·塔亚特合作的设计充分展示了艺术家的空间实验与设计师对身体和面料关系的思考所达成的一种动态结合。她的设计由艺术家来做时装图样,充分体现了这种动态的联系,也让她的设计成为未来主义的女性特质典范。模特的身体还有身穿的服装都被细细分割,不仅展现了它们的三个维度,又展示了它们的运动线条及其内在的现代性。

如果说普瓦雷与艺术的联袂借助了他在设计表达中对奢华与自由的渴望,那么对维奥内来说这成为她探寻处理与表现人体新方法的一部分。尽管这两位设计师的作品都复杂精细,它们还是被投入到大量的复制生产中。他们心中对生产商们大肆滥用自己作品的忧虑暴露出现代时装(当然,也包括艺术)固有的矛盾。正如特洛伊所指出的那样,面临风险的不只是艺术完整性的典范,复制行为也可能会破坏他们的生意,损害他们的利益。

鉴于艺术与时装逐渐深入商业世界,艺术家和设计师们不可避免地会把大批量生产的成衣视作跨界合作的新板块。这类合作项目使这两个领域之间的冲突以及它们与工业和经济之间的关系突显出来。它可能表现在认为艺术具有改变大众生活的强大力量的政治信条上,正如在俄国构成主义派设计师瓦瓦拉·史蒂潘诺娃1920年代的作品中表现的那样。当与她同时代的设计师们几乎都因时装的短暂性而避开它时,她却由衷地感觉到,虽然时装与资本主义和商业确实存在问题重重的关联,但它必然会变得更加合理,这与她对苏联时代"日常生活"的评价是一致的。她因而与她的构成主义派同侪们分道扬镳,开始明确提出:

> 认为时装将会被淘汰或者认为它是一种可有可无的

经济附属品其实大错特错。时装以一种人们完全可以理解的方式，呈现出主导某个特定时期的一套复杂的线条与形式——这就是整个时代的外部属性。

时装能够更为直接地联结更广泛的群体，这种能力让它一直都是那些想让自己的作品进入大众领域的艺术家们理想的工具。这大概遵循了普瓦雷在20世纪之初确立起来的先例，比如1950年代毕加索为美国一系列纺织品设计的那些妙趣横生的印花图案就体现了这一点，这些图案设计最终为许多时装设计师所采用，其中就有克莱尔·麦卡德尔。到了1980年代初，薇薇恩·韦斯特伍德与涂鸦艺术家凯斯·哈林的合作设计在精神上更加接近夏帕瑞丽跟艺术家们的联名合作系列。不管在哪个例子中，他们的协同创作都体现着一种相投的兴趣与意图，以韦斯特伍德与哈林为例，他们都喜欢街头文化，还喜欢挑战人们对身体的既有观念，这些都表达为装饰着艺术家画作的各式服装。

许多时装与艺术联名合作中最为核心的商业考量与消费理念，在20世纪末到21世纪初开始变得更加显而易见。川久保玲采用的严谨知识分子式设计方法当然毋庸置疑，但令人眼前一亮的还有她成功地调和了艺术追求、时尚还有消费三者之间潜在的令人担忧的关系。彼得·沃伦将日本设计师处理这种关系的方式与维也纳工坊的艺术家们进行类比，后者始终致力于把服装作为"大环境"的其中一部分来设计。这一环境涵盖了或许最为重要的零售空间，对CDG来说，它已经演变成了川久保玲设计美学的朝圣所，并成为从未中断的跨界联名合作点。一流的建筑师们,比如伦敦设计团队"未来系统"(Future Systems),为她在纽约、东京和巴黎设计了多家精品店。她的室内陈设模仿了标志性的现代主义艺术作品,比如她的华沙游击店中看似随意的设计布置,其实借鉴了包豪斯派设计师赫伯特·拜耶在1930年的法国室内设计师协会年展德国章节中呈现的开创性作品,即固定在墙面上的一模一样的椅子。

川久保玲和阿尼亚斯·B.等设计师一样,进一步利用了商业零售空间与画廊之间的模糊性,开始在自己的精品店里举办展览。在CDG东京店中,产品展示包括辛迪·雪曼的摄影作品,它们与时装通常的陈列相得益彰。类似的展览早已有之,比如纽约百货公司罗德泰勒在1920年代晚期就举办过一场装饰艺术展览,伦敦的塞尔福里奇百货1930年代也展出过亨利·摩尔的雕塑作品。不过,到了20世纪末它们之间的关系更为复杂,两个领域之间的关联也更牢固地确立起来,在艺术家与设计师们探讨身体与身份的作品中尤为如此。

21世纪之初,艺术与时装之间的关系依旧令人担忧,它同样展现了文化价值观以及人们潜意识中的各种欲望。艺术中的时装与时装中的艺术两者间的界限变得模糊,其各自展示的空间也变得不再泾渭分明。商店、画廊及博物馆运用类似的方法来展示并突出艺术与时装的消费,以及它们各自头戴的文化光环。路易威登出资为品牌与理查德·普林斯联名合作的2008春夏系列印花手袋举办的盛大派对就是一个极好的例证。这场派对在纽约古根海姆美术馆里举行,并选择了正在该馆展出的普林斯个展的最后一天晚上,引来了一些专业领域媒体的评论,评论探讨了商业赞助带来的诸多问题以及时装在美术馆中的地位。这个例子充分说明，虽然艺术与时装密不可分地联系着，但是当它们被太过紧密地捆绑的时候，两者在相互比较中既可能会获益，也可能会有所损失。

缪西娅·普拉达一直在积极地检验这些不同的观点。1993年，她专门建立普拉达基金会来支持并宣传艺术事业。她还请来雷姆·库哈斯等知名建筑师为自己设计标志性的"中心"（"epicentre"）店，这些店铺为她在店铺层展示时装设计的同时举办艺术展览提供了空间。其中就有她在纽约苏荷区门店展示的安德烈亚斯·古尔斯基巨幅摄影作品。古尔斯基作品中对消费文化的频繁抨击反倒给普拉达展出他的作品增添了一抹反讽的意味。于是，建筑师、艺术家和设计师们都呈现出一种心照不宣且自知的状态，他们创造着时装、艺术还有建筑，同时又不约而同地品评着自己的这些行为。

缪西娅·普拉达与时装及艺术之间的复杂关系，在她打造的名为"腰肢以下：缪西娅·普拉达、艺术与创造力"（*Waist Down: Miuccia Prada, Art and Creativity*）的裙装展览中表现得最为淋漓尽致，她用这次展览检视了自己过去时装系列中裙装设计的演变过程。展览由库哈斯的建筑团队设计，在全球的不同场地进行巡展，比如2005年的上海站就在和平饭店举办。展览运用了大量实验性的展示方式：有的裙子从天花板上悬垂下来，撑在特制的机械衣架上不停旋转；有的裙子平铺展开并覆上塑料外层，看上去宛如一只水母装饰物。普拉达的商业敏锐性与全球范围内的成功经营使得所有这些创新设计成为可能，而她在艺术及设计世界里的人脉则促使它们最终得以实现。

图8　2006年"腰肢以下"巡回展采用了充满创意的方式来展示普拉达裙装

然而，普拉达本人似乎格外钟情于这些关系的不确定性，同时对于这种不确定性如何联结起时装和艺术，她的认识也自相矛盾过。当展览于2006年移师她的纽约精品店时，普拉达对记者卡尔·斯旺森表示"商店本来就是艺术曾经的立足之地"，但紧接着又为她在这些"中心"店里的展览和其他作品的地位进行辩驳，她强调：

> 这是一个为实验而存在的地方。但把展览设在门店里并不是偶然。因为它从一开始就源自我们希望在店铺中引

入更多东西的想法,主要用来探讨我的作品。它其实是对作品的进一步阐释,与艺术全然无关。其存在只是为了让我们的门店更加有趣。

这一矛盾居于时装与艺术关系的核心。艺术家与时装设计师们的跨界合作能够产生有趣的结果,但双方也都会不安于大众对这类作品的接受程度。作为视觉文化的两个重要组成部分,时装与艺术始终表达着并且不断构建起诸如身体、美还有身份等各类观念。不过,艺术的商业的一面通过它与时装的紧密关系得到揭示,而时装又似乎利用艺术为自己注入严肃与庄重感。这些跨界合作体现出,每一种媒介都可能既信奉消费主义,又充满概念;既富有内涵,又关乎外在的展示。正是这些相通之处让时装与艺术走到了一起,并为它们之间的关系增添了有趣的张力。

第三章

产　业

1954年，英国人丹尼斯·尚德导演了一部短片《一条裙子的诞生》(*Birth of a Dress*)。影片开头取景于伦敦街头恣意陈设着时髦成衣的商店橱窗。随着镜头扫过这些橱窗外层锃亮的玻璃，旁白评论起英国女性极为丰富且触手可及的时装，并认为高级时装就是成衣设计的灵感来源。镜头紧接着特写了一件修身酒会礼服，裙身一侧荷叶边缀至裙摆，传达出1950年代晚装的韵味。片中介绍道，这条裙子首先由伦敦著名设计师迈克尔·谢拉德设计，经过改良后成了一件能够大批量生产的服装，使"街头普通女性"也能轻松购买。影片随后详细地展示了整个生产过程。时尚媒体通常选择回避服装制造这一产业化背景，而《一条裙子的诞生》赞美了英国的制造与设计奇迹，这些都参与到一条裙子的生产之中。这部影片由英国煤气委员会与西帕纺织公司赞助拍摄，揭开了一系列生产工厂的面纱，在那里用于服装制造的棉花被漂白以作备用。观众跟随镜头进入面料工厂设计织物印花的艺术家工作室，本片拍摄的是一款以炭笔描绘的传统英式玫瑰纹样。之后蚀刻工艺将印花图样转刻到滚筒上，科学实验室则开发出了苯胺染料（油气工业的副产品之一），然后工厂印制出大量面料，如此种种都在影片中得到了傲人的展示，以作为英格兰北部工业技艺与创造力的证明。

镜头接下来转到了迈克尔·谢拉德在伦敦梅菲尔富人区的精致沙龙，在这里迈克尔以这一印花面料为灵感，设计出一件晚礼服。以迈克尔的设计为基础，北安普顿工厂里的成衣设计师重新改良了这条裙装使其能够大批量生产。他们简化了设计，最终以三色印刷工艺推出一款时髦礼服，并通过一场时装秀向来自世界各地的买手进行了展示。通过镜头记录，影片让观众关注到女性身上时髦衣饰生产所必须历经的各个环节。这一设

图9　1954年影片《一条裙子的诞生》剧照，影片追踪了一条大批量销售的裙子从设计到制造的全过程

计离不开英国在高级时装设计和规模化时装生产中的成功,观众在镜头的带领下目睹了这些服饰是如何"与工业研究和科学发展的最新成果紧密相连的"。这部影片其实是战后一部反映英国工业发展、国家声望崛起、消费主义兴起的宣传片。它另辟蹊径,聚焦了时装的生产制造过程,将整个产业的各个方面都联结在一起,而通常它们都是以碎片呈现的:比如一件成品,设计师的概念,或者想要达到的一个目标。

正如《一条裙子的诞生》展示的那样,时装业包含了一系列相互交融的产业领域,这个领域的一端聚焦生产制造,另一端关注最新潮流的推进与传播。在生产者们忙于处理技术、劳动力问题,努力实现设计商业化的同时,媒体记者、时装秀制作人、市场营销者以及造型师们将时装打造成视觉盛宴,将流行趋势传达给消费者。服装已然被这些产业深深变革,从字面上说是通过生产过程,而隐喻意义上则是通过杂志与照片。所以,时装产业不仅生产了服装,更制造出一种丰富的视觉和物质文化,给人们创设新的意义、乐趣以及欲望。

安德鲁·戈德利、安妮·科尔申和拉斐尔·夏皮罗在他们关于这一产业发展的文章中指出,时装是建立在变化之上的。时装天生就是不稳定的、季节性的,时装产业的每一个方面都在不停地寻找调和这一不可预知性的方法。流行预测机构提前数年发布面料、主题等板块的流行趋势,用以引导、启发时装生产者们。时装品牌聘请富有经验的设计师,他们在自己对流行演变的直觉和标志性的个人设计之间取得平衡,从而推出成功的服装产品。时装秀制作人与造型师会将这些服装以最引人入胜的形式展示出来,以提升品牌形象,争夺报道版面,同时吸引商场店铺下单。店铺买手依靠自己对顾客特点、店铺零售形象的掌握采购最有可能畅销的服饰,也进一步强化他们所代表的零售商在时尚方面的可信度。最后,时尚杂志、高级时尚书刊等时尚媒体资源将会为时装制作广告、撰写时尚评论,竭力蛊惑、抓牢读者的心。

14世纪中叶开始,时装的发展便一直基于技术和工业上的突破,并得到了长期以来对传统小规模、密集劳动方法的依赖的调和,后者保有必要的灵活性以应对季节周期性需求变化。需要强调的是,时装产业也被消费者的需求驱动着。18世纪时,面料设计和潮流风格从年度变化转为季节性变化。着装者会依当季的流行趋势来改造自己的服装,借助裁剪和配饰打造全新的效果。正如贝弗利·勒米尔所写的,当富人们为昂贵的新款定制时装买单时,下层群体则将二手衣饰与17世纪的成衣搭配在一起。

很显然,时尚从来不仅仅是一个简单的效仿过程,无论是效仿贵族,还是后来的法国高级时装风格。虽然我们不能假设所有人都会,或者就此而言能够紧跟时尚,但消费者需求对这一产业的发展来说确实是一个影响深远的因素。从文艺复兴以来,对身份、美感以及在服装中获得乐趣(不管是触觉还是视觉上的)的渴望,都在其中发挥了作用。这一产业因而孕育出本土化、民族化和国际化的时装风格,生产者和营销者又不断努力迎合着多样化的欲望与需求。自18世纪英国年轻学徒们引导的地区性时尚潮流开始,他们通过向服装上增加装饰以脱颖而出,或者早至16世纪佛罗伦萨权贵身披的精工天鹅绒,时装业就包含了由交易员、分销商、推销者等构成的复杂链条。

时装产业的演进

文艺复兴时期的时装产业因纺织品的全球商贸往来而蓬勃发展,东西方商品相互自由交流。服装制作应用的技术逐渐成熟、优化,16世纪西班牙的裁剪教材使得更合身的剪裁成为可能。15世纪晚期,战争和贸易使得不同服饰风格在西方世界扩散,其中勃艮第宫廷的浅色风格占据主流,而深色系的西班牙风格则在接下来的一个世纪里不断传播。这种种潮流都是消费者对奢华和炫耀的欲望的一部分,等到17世纪路易十四对法国纺织品交易施行了制度规定,这种欲望开始变得正式化。这一举措不但巩固了已有好几个世纪历史的纺织品生产与全球贸易网络,法国最高统治者所采取的这些行动同时也认可了时装不仅能够塑造民族身份,还可以影响它的经济财富。于是,这一需要在不久后见证了巴黎高级时装公会的早期雏形于1868年成立,为的就是监督巴黎的高级时装产业。与之对应的是,新兴工业化国家也在持续不断地建设着自己的时装和制衣产业,墨西哥的生产力在1990年代的提升就是一个很好的例证。

17世纪见证了里昂的丰富面料、巴黎的奢侈品贸易、伦敦的成衣业日渐得到认可和巩固,它们凭借小规模的服装制作,依托专注传统技艺的小型工作室、家庭作坊快速发展。这些生产带动了本地富裕阶层和旅游观光者对时装的消费,它们也是成衣制造的早期尝试,在未来会给时装产业带来更广泛的影响,也就是让更多的人穿上服装,产生更多的经济利益,时装产业也最终发展成为一个重要的国际经贸门类和一股强大的文化力量。

军需生产为成衣产业发展提供了极大的动力。"三十年战争"（1618—1648）期间建立起了一支庞大的军队，军品工厂与外包制衣工坊一起为军人生产制服。18世纪及后来的拿破仑战争时期，这类生产进一步扩大。早期的成衣生产主要是毫无个性的服装，例如给海员制作宽大的工作服，裤腿肥大通常是他们衣着的特点，还有为奴隶生产的基本服装。尽管本身并非时装制造，这些生产活动还是为即将出现的成衣产业提供了必要的先决条件。

美国作为独立国家的崛起在成衣产业的发展中扮演了一个至关重要的角色。1812年，美国军队服装公司在费城开业，成为最早一批成衣生产商之一。伴随美国内战带来的巨大的军服需求，还有淘金热给李维斯带来的牛仔生意，一个基于生产方式和服装尺码标准化程度更高的产业兴起了。克劳迪娅·基德韦尔在她的书中指出，19世纪末人们对成衣的态度出现了一种与之对应的变化。成衣不再被当作拮据和底层身份的表征。城市化不断推进，城市工人和居民需要买得起的服装，并且要"跟主流时装看上去没有多大差别"。城市之中更多时尚潮流随处可见，人们期望自己在人潮之中与众不同的愿望，成为驱动成衣产业的另一股力量。

需求总是和创新紧密相连。珍妮纺纱机（约1764年）提高了面料生产速度，提花织机（1801）使更复杂的织物设计成为可能。然而，正是合理的尺码标准体系的形成使得高效的大规模生产成为可能，也使得时装产业从19世纪中叶开始更加发展壮大。举个例子，根据菲利普·佩罗的著述，截至1847年巴黎共有233家成衣生产商，雇员规模达7 000人；而在英国，据1851年的调查，服装贸易雇用的女性数量位列第二，仅次于家政服务。从这一点看，女士成衣产业同样也在发展，与早期男士成衣的情况类似，女士成衣主要是斗篷这类的宽松服装。

　　胜家公司于1851年投入市场的缝纫机，有时被看作成衣行业发展的革命性因素。然而，直到1879年蒸汽或燃气驱动的缝纫机往返摆梭被发明出来，服装生产的速度和简易性才得到大幅度提升。安德鲁·戈德利曾有记录，一名熟练的裁缝每分钟能够缝制35针。但到了1880年，动力驱动式缝纫机能够每分钟完成2 000针，而到1900年这一速度已经攀升到每分钟4 000针。技术的进一步革新，例如裁剪和熨烫技术的改良，在极大缩短生产时间的同时也减轻了生产者和消费者的成本负担。

　　1880年代从俄国种族迫害之下出逃的大量移民，给英美两国的成衣产业增添了更多动力，犹太裁缝和企业家也成为时装产业发展的重要力量。例如，伊莱亚斯·摩西（Elias Moses）在广告中号称自己是"伦敦第一制衣……开创了成衣新纪元"，而后又宣称"当前的制衣速度和火车一样迅捷"。摩西把自身行业效率飙升的技术和更为迅猛的交通方式类比，确实再贴切不过。火车路网不仅加快了贸易和分销，更打开了传播的可能性，将时装潮流扩散到各个阶层甚至不同国家之间。

　　从19世纪中叶妇女们头戴的用来遮挡海边阳光的黑纱，到越来越受欢迎的舒适的男式西装便服，旅行度假服饰及运动休闲时装促进了成衣产业的蒸蒸日上。19世纪最后四分之一的日子里，女性开始进入白领职场，她们迫切需要适合公共场合的新款服装。作为女性着装原型模式的"裁缝定制"也在1880年代得到了新的发展。穿上了女士衬衫的新女性们代表了一种全新的风格，在此背后19世纪末不同性别群体中众多的新时装风格蓬勃兴起。确实,美国制造的"仿男士女衬衫"在1890年代之初就已经风靡,证明了消费者需求与供应商创新之间的紧密关联,它不断驱动着时装产业向前发展。

如果说18世纪见证了点燃大众对时装欲望的西方消费文化的发展,那么19世纪则将这种对新奇与感官体验的热爱转变为遍及全球的视觉和商贸狂潮。发明家们接连开发出大批量生产的裙衬、束身内衣、裙撑等一个又一个专利产品,用最新的技术重塑着女性的身体;橡胶、赛璐珞材质的出现,给渴望拥有精致绅士风度的年轻男性提供了既可以轻松负担又容易打理的洁白领子与袖口;苯胺染料的出现意味着织物可以大胆地将科学革新与时尚融合起来。

成衣产业加速发展的同时,高级时装业也越来越多地习得了商业往来中的博弈伎俩。露西尔和沃思等代表性设计师以及主要的几家百货公司所使用的宣传策略,尤其是时装秀,发挥了巨大作用,将各种流行风格的华美形象广泛传播。这些展示活动在市场的各个层级传播并造成影响,也给想要将最新潮流改造为符合自身价位的生产商们提供了模仿的样板。美国买家尤其热衷于挖掘高级时装真品气质中的商业潜力。他们花钱参加各场时装秀,与卖家达成协议后采购一定数目的服饰,然后根据这些服饰生产限定批量的复制品。如在17世纪之时,"巴黎"就是奢华的同义词,这个城市的名字出现在各式广告、专题报道之中,被世界各地无数的商铺和品牌放在自己的名号里,以作为时尚可信性的标志。巴黎代表了优雅和旧世界的奢华,当其他城市努力打造能在国内国际市场上畅销的特色时尚风格时,巴黎也为它们的服装产业提供了一种发展范式。

至19世纪末,时尚作为服装产业的一股主要推动力不断勃兴,它把流行服饰带给了更广泛的人群。时装一方面使得人们得以构建新的身份;另一方面,产业发展的背后是对以女性和移民群体为主的产业工人的剥削。血汗工厂是这一充满活力的现代化进程的阴影,始终挥之不去。自1860年代起,行业丑闻不断爆出,确保如期交货及低廉零售价格的逼仄空间、长时间劳动和极微薄薪水的故事一次次震惊了政府和社会民众。关于生产道德的讨论带来了大规模的工会联合,20世纪早期,限制最低薪酬的法律也得到推行。尽管在剥削劳动力问题上臭名昭彰的是致力于大规模、标准化服饰生产的服装业,时装引发的一系列争议从未中断。维多利亚时代缝制定制礼服的那些孱弱年轻女子的形象,在今天被替换成各路信息曝光中服装品牌在亚洲和南美洲雇用的童工。

时装生产商们自古就需要保持与市场的密切联系,并针对消费者对某一特定流行趋势的需求迅速做出反应,今天发达的信息网络让服装生产装配可以分包到越来越远的地方。随着20世纪渐渐远去,科技手段使得每一个服装款式的销量数据能在各个店铺的各个收银机上进行整理检点,进而能迅速向工厂下单。运输方式和分销渠道的改进进一步加速了这一过程,极大地为一批成功的国际服装巨头们增加了便利,比如瑞典的海恩斯莫里斯(H&M)、西班牙的飒拉(Zara)。这些公司能够快速响应设计师的时装系列,同时密切关注着街头流行的新兴趋势,以复制最新的高级时装,有时候甚至能抢在高级时装上市之前先推出自己的产品。这种模式同样意味着工人的工作环境更加难以保障,引发了针对高街品牌的种种指控,盖璞(Gap)的遭遇就是个例子。

现代时装产业的结构在1930年代已经确立。随着20世纪逝去，这一产业被冠上"快时尚"的名头，从前按季发布的节奏已经被打破，服饰产品如今能够不间断地出新，持续供应给各个高街时尚零售商。1920年代的高速增长为这套系统奠定了基础。那个十年见证了更大规模的投资、更大范围的国际交流，同时越来越多的迹象表明，时尚，而不是质量和功能，能够赋予服装甚至汽车更多的卖点。"大萧条"开始后，人员和财力的大量削减让人们更注重精简有效的工业生产，建设国内市场，同时在全球范围内寻求新的区域以作为目标市场，巴黎的高级时装设计师和美国的成衣生产商们，都把南美地区视为新客户的重要潜在来源地。

战后时期，消费市场和国内工业得到了进一步的巩固。在美国的支持和商业知识的帮助下，意大利和日本发展出了各自的时装产业，在时尚服装与衣橱基本款之间取得了平衡。确实，这一结合非常重要，泰莉·艾金斯认为它对于延续一个品牌的商业生命至关重要。她断言美国设计师艾萨克·麦兹拉西1998年不得不关闭自己的同名品牌业务，就是因为他把全部精力都放在了时尚服装上，而忽略了经典款式的需求。

这再一次证明了时装产业的善变，设计师和生产商必须考虑周全才能增加并稳定自己的市场份额。这一点体现在高级时装设计师品牌授权生产、开发的成衣产品线，以及20世纪晚期出现的成衣品牌衍生的副线上，比如高缇耶童装（Junior Gaultier）和唐可娜儿（DKNY）。这些系列依托设计师们通过各自主线打造出的光环，并借助价位更亲民、款式更基础的服饰不断拓展消费人群。

时装产业对外部投资和其他资本保障途径的需求在20世纪得到解决。英国博尔顿(Burton)男装生产并零售自有设计,使得供需两端间形成紧密的关联互动,也使公司在1929年得以上市。从1950年代晚期开始,法国时尚品牌开始上市。1980年代以来,奢侈品巨头,比如路威酩轩,旗下囊括莫杰(Marc Jacobs)、路易威登(Louis Vuitton)、纪梵希(Givenchy)、凯卓(Kenzo)以及璞琪(Emilio Pucci)等品牌,将年轻品牌与老牌时装店汇合,以保证其时尚可信性,同时通过将利润分散在名酒、香水、腕表及时尚品牌等众多领域以抵御经营风险。

然而,时装产业依然有一个重要门类始终保留着几个世纪以来的规模极小、劳动密集型的经营模式。主要分布在伦敦东区的许多工作坊就是这一类型的缩影,年轻设计师加勒斯·普、克里斯多弗·凯恩以及马里奥斯·施瓦布只雇用了极少量的助理来协助他们完成自己的服装系列。他们沿袭着英国设计师们自1960年代就树立起的传统,作品出色的时尚设计引起了媒体的兴趣,影响力随之远播世界各地。

时装宣传推广方式的发展

时尚媒体与广告行业紧跟着服饰生产和设计的脚步不断发展,传播着最新流行趋势,并通过影像和文字构建起种种时尚典范。毫无疑问,媒体报道能够让设计师迅速蹿红,就像普、凯恩和施瓦布那样,但它也有可能侵蚀长期发展的根基。如果设计师在职业生涯之初就过快地斩获盛名,而此时的他们还没有找到有力的资金支持,不具备与订单需求相匹配的生产能力，他们的事业便难以成长。尽管如此，媒体报道仍然被认为对于树立品牌，并最终找到来源可靠的资金投资至关重要。这样的矛盾在伦敦时装周表现得尤为明显，以中央圣马丁艺术与设计学院为代表的艺术院校不断培养出大批极有才华的设计师，但基础条件匮乏、政府投资不足又让他们的设计事业不堪一击。

20世纪下半叶，循环举办的季节性国际时装秀逐渐主导了时装产业。它们为设计师和生产商提供了一个平台，在这里他们能够按自己希望的方式展示自己的设计系列，而不是透过杂志报道的滤镜来展现。时装秀集合了全球销售终端的买手、消费高级定制时装的富有个人客户等，随着它们的发展壮大，媒体从业者和摄影师也参与进来。从19世纪末规模极小的高级时装沙龙展示开始，时装秀逐渐形成自己独特的视觉语言，不仅包括模特的动作和仪态，灯光与音乐的配合，还有用来传达各个品牌标志和追求的那些越来越精美的现场表演。

直到1990年代，时装秀的内容都是经过媒体筛选编辑后才向大众传播，不管是报纸杂志，还是后来法国时尚电视台这样的电视频道。然而到了20世纪晚期，互联网向公众提供了能够接触到未经编辑内容的途径，这些内容有时会在设计师的网站上同步直播。当下的即时性有可能打破设计师与生产商，时尚媒体、零售商和潜在客户之间原本的稳定关系。它让设计师的作品以完全不经媒体加工的本来面貌呈现给消费者，消费者可以直接购买秀中的服饰，而这些款式时尚杂志和店铺买手们则不一定会相中。

出版、广播再到现在的互联网媒体，它们构成的国际化网络凭借着戏剧化的意象创造时尚内涵,几个世纪以来不断演变。文艺复兴时期,无论本地城镇还是国外的贸易和商旅,都会带来新的时尚资讯。讽刺文同时嘲笑和赞美着时尚。一流的制衣商们设法通过散发玩偶以传播流行趋势,而这些玩偶身着的都是最新款式的正装和便服。书信往来为传播最新款式提供了一条非正式途径。确实,相比小说,简·奥斯汀在她与姐妹的通信中留下了更多的时尚信息,详细地写到了她们帽子上新的装饰、新购置的裙子。这种近似名人逸事形态的时尚传播,在今天的网络博文中继续存在,也体现在一些内容风格个人化的小刊物里,例如专注复古风格和自制时装的杂志《邂逅》(*Cheap Date*)。

17世纪,更加正式的传播方式不断演变,包括自早期展示不同国家服装的画册演化出的不定期出版的时尚杂志。不过,第一本定期发行的时尚杂志直到1770年代才正式出版。《淑女杂志》(*The Lady's Magazine*)建立了整个时尚报道和形象塑造的行业。最令人印象深刻的大概是这本杂志的开本,到21世纪的今天它仍被当作范本。18和19世纪的时尚杂志的内容,既有充满谈资的社交事件,文中事无巨细地描述名流们的装束,也有给读者们的护肤化妆和着装建议,还有小说以及巴黎顶级高级时装设计师和裁缝师的各类新闻。在如何选择合适的时装、美妆,如何举止得当等问题上,时尚杂志把说教式的文章和"姐妹"般的建议融为一体,颇为有力。它们构建起"女性"的典范,不管是忠于家庭生活的强烈道德观还是对性别身份的前卫挑战,前者体现在19世纪中叶《英国女性居家杂志》(*Englishwoman's Domestic Magazine*)对家务和服饰纸样的建议中,后者则出现在

1980年代中期的《面孔》(The Face)杂志里。

这些杂志很早开始就通过广告和更为隐秘的促销链跟时装工坊和生产商们保持着密切的关系。1870年代,《米拉衣装志》(Myra's Journal of Dress and Fashion)成为社论式广告的先驱,它把广告和专题评论内容糅合在一起,在刊载玛丽·古博夫人文章的同时配载了关于她的时装店的广告图片和评论内容。这种关系在20世纪进一步发展。1930年代,埃莉诺·兰伯特就是最早把公关技巧应用到时装产业的那批人之一,他们认识到了多元化推广的可能。这样,媒体代表开始游说将各大品牌投放入专题文章和图片中,通过增加时尚杂志的声望来确保已有广告版面的传播效果。兰伯特还鼓励自己代理的电影明星穿着与她稳定合作的设计师的作品。1950年代,她将运动服装设计师克莱尔·麦卡德尔的新款太阳镜系列送给著名影星琼·克劳馥,兰伯特清楚地知道克劳馥佩戴麦卡德尔作品的照片会成为设计师产品的代言,同时也会提升明星自己的时尚影响力。

类似的互惠跨界合作巩固了时装产业的基础。19世纪晚期,伦敦的一流高级时装设计师,比如露西尔,纷纷为舞台上的顶级女主角们提供服装,以此获得免费的宣传,增加自己产品的曝光度。这种操作继续发展,设计师们开始为电影设计戏服,比如纪梵希为奥黛丽·赫本1961年的电影《蒂凡尼的早餐》设计的那件标志性的高定礼服,又比如让-保罗·高缇耶1997年为电影《第五元素》(The Fifth Element)设计的卡通化的前卫科幻戏装。最关键的是,演员和名流们会在他们的私人生活中穿着某些时装,让人们更加认定某位特定设计师的作品与名流们的生活密不可分。美国电影艺术与科学学院奖①这类全球关注的盛事来临时,设计师们则要使出浑身解数争夺为明星提供服装的机会。《问候》(Hello)这类杂志会追随早期好莱坞影迷杂志的方向,进一步模糊明星公众与私人状态的边界,它们拍摄名人在家中的照片,并在旁边罗列出明星们身着所有服饰的品牌信息。

时装产业和它这些"盟友"间方方面面的相互依存,常被诟病为是在创造千篇一律的可认同身份。尽管这些批评一定程度上是中肯的,因为苗条、白皙和年轻无疑是主流的形象,但时尚同时也不停地探索着许多边界。时尚和潮流杂志本身就是孕育它们文化的一部分,因而反映了人们对种族、阶层和性别等更多元的态度。时尚杂志代表新事物的定位,以及它们对一流作者、形象策划人的强大吸引力,同样意味着它们能够提出全新的个性,并为读者创造出一次逃离日常生活的机会。1930年代,美国版《时尚》(Vogue)倡导富有活力的现代女性形象,将暗示自由和激情的夸张的女飞行员图片与上班着装的实用建议刊载在一起。1960年代,英国刊物《都会男士》(Man About Town)会为它给男性读者的时尚生活建议配上身着精致利落西装的男士图片,照片背景是鲜明的都市场景。1990年代后期俄罗斯版《时尚》中极尽奢华的高级时装则成为让人们暂时忘掉经济危机烦恼的梦幻之境。

每一种出版物都形成了自己的风格,竭力吸引着读者,并为读者们的时尚地位提供了一种标记。20世纪早期,《皇后》(The Queen)杂志代表了优雅的精英范儿;1930年代,主编卡

---

① 又名"奥斯卡金像奖"。——编注

梅尔·斯诺和艺术顾问阿列克谢·布罗多维奇打造出高级时尚杂志《哈珀芭莎》(*Harper's Bazaar*)，以富有冲击力的文字、图片和插画节奏夸张地呈现出现代主义优雅形象；而1990年代在比利时安特卫普出版的《A杂志》(*A Magazine*)，请来马丁·马吉拉等一众前卫设计师作为各期杂志的"策展人"。商业出版物提供了近似于多数报纸杂志报道的梦幻气质的替代品，也在联结时尚各个不同元素上发挥了同样重要的作用。19世纪出版的《裁缝与裁工：时装行业刊物与索引》(*The Tailor and Cutter: A Trade Journal and Index of Fashion*)刊载了许多实用信息和技术讨论文章。1990年代之后，以世界全球潮流互联网（WGSN.com）为代表的网站媒体，已经能够海量整合世界各地分站的行业顾问对未来潮流的预测，以及关于全球各个城市的街头时尚的报道，而这些资讯让时装产业能够即时获取新兴潮流和发展信息。

时尚杂志创造的图像与文字、人体与服饰、社论与广告等的拼贴，为读者制造了一个能够遁入其中的纷繁空间。它们建立起一座视觉消费的王国，在那里甚至是纸张也力求能给读者带来多重感官体验，不管是法国著名杂志《她》(*Elle*)平滑而有光泽的用纸，还是《新杂志》(*Another Magazine*)的纹理材质插页。虽然时尚杂志总是迅速过时，它们却是同时代文化和社会发展的记录，并将时装产业的商业规律和它自身在全球视觉文化中无形的推动作用统一到了一起。杂志不只是报道各类时尚内容，在很多人眼中，它们本身就是时尚。插画和摄影艺术为服饰增加的含义有时也将它们转化为时尚。一面是服装的日常现实性，一面是插画家妙笔生花或时尚大片点石成金所制造出的幻象,种种新观念在这个层面之间不断地产生影响。这些新观念渗入当下的风俗,但又不断地突破固有观念以提出增强的现实或超现实主义叙事方式。

  在下面这幅19世纪早期时装图样中,画师简化了他的速写线条,呼应了时髦廓形简洁的特质。图样选取中心人物的背面视角,画面焦点落在女士身上仿古典式的垂坠打褶裙装,并以她围裹的明艳红色长披巾强调了这一形象。男士外套身后收窄的燕尾也在他以古典"裸体"为灵感的裸色马裤对比下凸显出来。画中还有许多时尚细节,如男人们极为时髦的鬓角,席地坐着的女士头上猩红色的小帽,都被置于这幅插画的叙事之中。时装图样给服饰增加了情绪和语境,原本仅在下单时作为制衣师或

图10　1802年的一幅时装图样,展现了这一时期仿古典式的时装风格

裁缝们模板的简易插画的原始信息开始得到丰富和提升。图样创设的环境能够营造出一种自在闲适的优雅感觉,把时装和更多样的风尚联结在一起,在这幅画中画师选择了当时人们对热气球的迷恋。

发端于19世纪中叶的时装摄影也发挥了类似功能,增加了在真人身上展示服装这一元素。如果说时装生产在努力平衡时尚难以预测的天性,那么时装影像则在赞美时尚的模糊性。图像在时装的建构中扮演了核心的角色,它展示了不同风格的时装上身后的效果,并将与特定服装相关的动作与姿势编入目录。

美国摄影师托尼·弗里塞尔这幅1947年的摄影作品说明了简单普通的日常衣着是如何通过图像得到改变的。弗里塞尔没有将这件网球服展示在它通常会出现的球场环境里,而是将模特置于一片突出的山地景色中。自然光线里,裙子亮白色的面料光彩熠熠,利落的廓形在阳光中进一步凸显。在观者眼中,模特的身份无法识别。她将头转向山景的一侧,身体姿势强化了她健康运动的形象,但又表现得十分自然。阳台的曲线将模特和线条流畅的现代建筑联系起来,为她营造出一个传递着自然与人工两种高级质感的场景。时尚编辑对模特的挑选,对拍摄造型的设计——干净的橡胶底帆布鞋和及踝短袜、清爽的发束、身边随手放置的羊毛衫,都加强了弗里塞尔布景和构图投射出的漫不经心的悠然。这样,成衣因此情此景被赋予了一种其原本缺乏的时髦气势的光彩。

这些相互交织的产业周旋于生产商和消费者之间,因而创造出各种各样的"时尚"显现点。这些点都是循序渐进、不断累积的。约翰·加利亚诺所受过的时尚训练、他的经验和个人

图11 托尼·弗里塞尔1947年的时装摄影作品,身着网球服的模特置身于一片突出的风景之中

直觉使得他最初的设计中就包含了未来的风尚,而后经过不断演化,这些风格得到不断加强。与他一起在迪奥高级定制工作室共事的那群技艺娴熟的能工巧匠们进一步促成了高级时装时尚可信性的传统,这一传统已经薪火相传了几个世纪。通过精心设计和布置的环境、对模特和造型的戏剧化调度,加利亚诺用自己的时装秀向时尚界的各路人士展示着自己的风格主张。时尚媒体们秀后便开始强化并且可能重新阐释加利亚诺的时尚风

格，采取的方式是刊发核心潮流的描述文章，并将加利亚诺的设计作品与他的同侪们进行对比。无论是广告和报道照片，还是零售终端和橱窗陈列，都将加利亚诺的作品认定为时装，并提出多种方式以激发人们对如何穿着这些时装的想象。

我们很难确定服装具体在哪一个时间节点摇身一变成了时装。对于加利亚诺，或是更早期像20世纪中叶的巴伦西亚加这样的高级时装设计师们来说，时尚风格的诞生一方面要通过他们的设计实践，另一方面也仰仗他们的设计作品被传达给大众时所使用的一系列宣传和广告攻势。1930年代至今的成衣和高街时尚产品也是一种类似的综合体，它们既有逐渐确立起来的时尚可信性，又有时尚媒体的验证，还有通过衣饰给人带来丰富灵感启发的无形能力，这将服装和身体的典范与当代文化的其他领域联系在了一起。

第四章

购　物

2007年CDG在华沙开出一家新的游击店[1]。根据计划这间店只会在这里运营一年，以作为品牌类似的"快闪"店项目的一部分。第一家游击店2004年在东柏林开业，接下来品牌又在巴塞罗那和新加坡开设了相似的短期精品店。每间店都独具个性，又与它所在的环境相协调。华沙店完整保留了选址原本的苏联时代果菜店外观：绿色的瓷砖，不平整的抹灰，粗糙墙面上还有家具配件扯掉后留下的痕迹。整个空间的美学延伸到了"陈列橱柜"，这些柜子都是苏联时期的家具，被安装后用来放置品牌的产品。它们杂乱无章地贴在墙上；抽屉被拉扯出来，歪歪斜斜，半开着露出里面晶莹的香水瓶；破损的椅子从天花板上悬垂下来，破旧的坐垫上摇摇欲坠地摆放着各式鞋子；服装悬挂在光秃秃的金属横杆上；灯饰上垂下的电线拧作一团，盘踞在地板上，一半隐没在堆叠起来的家具之下。

---

[1] 一种在商业繁华区设置临时店铺，于较短时间内推广品牌的商业模式。——编注

图12　2008年CDG华沙游击店内部设计，看上去如同一场现代主义家具展

整个布置营造出一种被遗弃的贮藏室的效果，店主似乎匆忙逃离，只留下满屋的服装配饰。这种氛围是店铺地理位置与历史背景的一种象征——共产主义被前苏联阵营国家抛弃，资本主义即将取而代之。这样的变革也让人们改变了曾经只买需要的，或者说只能有什么买什么的消费习惯，转而开始在丰富的商品选择中消费自己梦寐以求的东西。店铺打造出的潦草感同时也呼应了游击店的本质，即它会突然占据某一城市空间。实际上这已经是品牌在华沙的第三个化身，第一次尝试是2005年在一座桥下的废弃通道里完成的。

尽管这些店铺看似偶然又毫无章法，它们却是CDG在时装零售业保持自己前沿地位的精心策略的一部分。一部分店铺只营业几天，有的则持续一年；所有的店铺均未进行广告宣传，除了以电子邮件告知老顾客,或者在开设当地张贴几幅海报,而最关键的是靠人们口口相传。这些过程仿效了亚文化的传播效果,只需告知小圈子内的舆论制造者,而这一群体其实早已认可品牌在时尚行业里代表前卫风格和设计的先驱者地位。游击店为自己的产品营造出一种独有、神秘又刺激的氛围。这种氛围促进了一种感觉,即来访者有知情的特权,让他们认为在这里购物也是参与了一场半私密的活动。通过强化欲望、生活方式和个人身份,品牌以此切入了21世纪早期高级时装消费主义中的关键元素。照此,这家游击店再次如各种街头文化一般,在彰显个性的同时又表明了自己属于某一群体。它主张购物其实是一种体验,对这间店来说,人在其中犹如参观一家小型画廊。重要的是,它以一种贯穿品牌知识分子精神特质的方式树立品牌形象。CDG显然拒绝了许多时尚广告与销售行为中那些过度且颓废的东西,同时针对品牌核心客户拥有一套精明的营销策略,也不断吸引着充满好奇的"路人"。

　　1980年代开始,设计师川久保玲就开设了一系列极具创新性的店铺。她早期的精品店中节约而格局紧凑的空间沿袭了传统和服店的美学理念,将服装产品整齐叠放在货架上。除此之外还通过只陈列少量产品来创造出一种虔诚的感觉,让购物者能够关注产品的细节和整体包装。她的竞争品牌以及像盖璞和贝纳通(Benetton)这样的高街品牌仿效了这种做法,也开始运用木质地板、纯白墙面,把毛衣成堆叠放在货架上,并精心布置强化空间感和利落线条的挂衣杆。

　　川久保玲和她的丈夫阿德里安·约菲2004年在伦敦开张的"丹佛街集市"则另辟蹊径,在整栋建筑物中的许多不同空间中精心展示各种时装和设计品牌。在某一层,更衣室被设置在一个巨大的镀金鸟笼之中;另一层上,服装却和花卉绿植、园艺工具组合在一起。川久保玲为"丹佛街集市"赋予了灵活性和多样性的概念,她在店铺的网站上写道:

> 我想创造一种集市,在那里不同领域的不同创造者能够聚集、相遇在一种持续的美好而又混乱的氛围中:共有强烈个人审美的不同灵魂在这里交汇,齐聚一堂。

这一氛围恰如一个19世纪集市的现代版本,充斥着一大批不断变化的独家时装产品线和不拘一格的物件。

跟CDG更为稳定的精品门店比起来,"丹佛街集市"这样的经营模式强调了现代零售业的多样化和灵活性。在饱和的市场环境里,所有的设计师和时装品牌都必须突出自己的个人特征才能建立稳固的消费群体。CDG代表了这一经营模式的前沿,它所采取的这些方法令人不禁想起更早年代的行业前辈,不管是19世纪清楚一定要给商品制造视觉冲击力的百货商场经营者们,还是20世纪早期将自己的沙龙打造成反映个人服装风格的私密感官空间的高级时装设计师们。

零售业的发展

文艺复兴时期,纺织品与镶边配饰仍旧从集市和众多流动商贩手中采买,这一传统延续了几个世纪。蕾丝、缎带以及其他配件要么被带到乡间四处兜售,要么批发给本地商店。规模稍大的村子里会有贩卖羊毛和其他材料的布商,城镇里会有销售高级丝绸和羊毛的女帽店。本地裁缝和鞋匠则会生产衣服和配饰，一件衣服的各个组成部分需要从不同的商店采购来再由匠人制作，因此，购置服装极有可能是一个漫长的过程。各个国家的购买模式也有所不同。在英国，为了买到更时髦的服装，人们经常前往附近市镇或城市。然而在意大利，地方各自为政、地理分离破碎的境况使得不同区域间的差异更为显著，因而造就了各个村庄中种类更加丰富的店铺。

全球纺织品贸易已发展成形上千年，国际商路跨越亚洲、中东直达欧陆。人们举办大型市集来向商贾和小贩购买及销售面料，他们会不断流动，前往布鲁日或日内瓦，或者去每年举办三次集市的莱比锡，又或者前往利兹的布里格市场。17世纪时，英国及荷兰东印度公司强化了与亚洲之间的商贸联系。到18世纪中叶，诸如印度棉布等已经成为日常面料。当时印度棉布相当时兴，更重要的是它既便宜又耐水洗，因而大幅提升了各个阶层人们的衣着清洁程度。得益于航海运输的改进，这些商品能够跨越全球运送。同时，对于时髦面料的需求激增，因为越来越多的人希望自己能够衣着时髦体面，能够符合当代社会外表和行为举止的典范。这些东印度公司用不断推新变化的织物设计，国外进口的丝绸、棉布、印花布等满足着人们的欲望。商人们鼓励时尚先锋们穿着自己的最新商品参加会被时尚杂志报道的时髦社交活动，以此来推广新的时装。伍德拉夫·D.史密斯曾记述了东印度公司如何安排印度工匠开发更多热门设计，然后随着时尚从巴黎传播出去而将这些时装卖遍欧陆。正如丹尼尔·罗什在提到法国裙装变化时所说的那样，到18世纪末，大体上购物者可选择商品的门类远为丰富，"但

图13　伦敦街头集市的二手衣物贸易已经持续了几个世纪

是一切与外表展示相关的东西,无论是社会还是个人需要,都增加得更多"。

纺织品和服装一度较为昂贵,被当作仆人月例的一部分来发放,并在家人手里一代代留传下来,也会在二手商店、集市一环接一环地出售,直到它们变成破布或者再生成纸张为止。随着18世纪耕作方式得到改进,财富分配有所提升,更多人想要购买时尚服饰,起码能让自己在礼拜日穿得最为体面。店主们开始在陈列展示产品以及服务顾客上花费更多时间。到了1780年代,平板玻璃橱窗的使用让陈列更加引人注目,店内的装饰陈设也开始愈发精致起来。时尚购物已然逐步塑造起城市的地理。伦敦的考文特花园成为第一个时尚郊区,伊尼戈·琼斯操刀设计的广场上进驻了各类布料商和制帽店,这些店铺在1666年的"伦敦大火"之后搬到了城西。在巴黎,皇家宫殿完成改建,某种意义上成为第一家专门打造的购物中心,沿着宫殿花园的四周,一排排小型商铺和咖啡馆延展环绕。广告和营销手段也在同步发展。传单上夸耀着某家店的各色成衣服饰,或是丰富的面料选择;时尚杂志里刊登着对最新款式的细致描述和精美插画,企业生产商和销售员们则极力鼓动时尚引领者们穿着自家产品出现在公众眼中。自文艺复兴起,购物与人们逐渐增强的个性意识一道携手前进。时髦裙装提供了在视觉上表达个性的手段,而清楚时装应该在哪儿买、怎么买则是达成目标的关键。托拜厄斯·斯摩莱特等小说家们讽刺了人们试图打扮得富有吸引力、时髦,同时试图使自己看起来高于自己的地位的行为。斯摩莱特察觉到了日益发展的消费文化,而这一文化在接下来的世纪里将繁荣兴旺。

购物的发展

1800年代早期,小型的专卖店仍旧十分重要,不过在大型公司兴起后大范围的商品和服务才聚集起来,这预示着购物将进入一个全新的时代。阿里斯蒂德·布西科1838年在巴黎创办乐蓬马歇商店,到1852年时它已然演变成一家百货公司。它把面料、男士服饰还有其他时尚产品集合在一起,并且通过在店内设置餐厅为购物引入了一项强烈的社会元素。布西科开发出多种顾客服务,这进一步促使人们意识到店铺员工与顾客间关系的改变,以及顾客与使用店铺之间关系的变革。他固定了价格,并在所有商品上进行标注,这样一来就省掉了讨价还价的必要,同时他也允许退换货。乐蓬马歇百货是最早的一批百货公司之一,另外还包括曼彻斯特的肯德尔·米尔恩,它由1831年的一个集市演变而来，以及纽约的斯图尔特，它由1823年的一间小布料店逐步转型而成，到1863年时已在纽约百老汇这一主要时装购物区稳坐头把交椅。这些百货商店演变出越来越成熟复杂的销售技巧。店铺欢迎顾客在店内随意浏览，循着精心设计的路线穿过店铺各楼层，走进店内的咖啡馆和餐厅，或是驻足观看店中安排的娱乐活动。购物首次变成了一种休闲的追求，专注于在一个时髦而令人安心的环境之中消遣时光，当然还希望顾客能花费金钱。

女性作为百货公司的主要目标，常常被它们精心布置的橱窗陈设中突显高级面料的光效、货品的丰富色彩和质感深深吸引，不觉间踏进这些精美的建筑。从前的中产阶级和上层阶级女性是不可能单独购物的。即便有女佣或随从的陪侍，某些街道在一天中的特定时段也在她们的活动边界以外。比如，伦敦邦德街的店铺主要是面向绅士们的，女士如果下午前往的话会被认为非常不得体。这些小心谨慎的礼法规定被百货公司慢慢侵蚀，它们鼓励女性在那里进行社交往来、任意走动浏览，苏珊·波特·本森曾引用波士顿一位店主爱德华·法林的话，称其为"没有亚当的伊甸园"。这不仅赋予了女性更大的自由，更塑造了作为消费者的她们。埃丽卡·拉帕波特以一种模棱两可的表述描绘了这一改变。维多利亚时代人们心中理想的女人应该一心系在家人和家庭上。女性顾客出门为她们的孩子和丈夫采买东西可以被看作是在关注这些家务事，另外她们也会给自己购买能够彰显她们家世身份和品位的时装。然而，外出购物同时意味着离开家中的私密环境，前往城市中心地带，置身先前男性主导的公共领域之中。购物同样更关注人的感官体验，而非那些更加高尚的女性消遣活动。用拉帕波特的话来说，这是城市作为"享乐之地"发展过程的一部分，在其中"购物者被命名为享乐的追寻者，由她对商品、场景还有公共生活的渴求来下定义"。时装因此给人带来了一种矛盾的体验。购买服装、饰品和男士服饰用品能让女性在19世纪不断发展的城市场景里占据一片新的空间，同时又潜移默化地引导她们接受一种注重矫饰和欲望的生活方式。店家们努力让自己的陈设尽显诱惑力，煽动女人们放任自我在他们的店面之内消磨整日，或是在大型市镇和城市里集聚的各类店铺之间流连。

每一家店铺都确立起自己独有的特性，志在招徕被自家风格和琳琅商品吸引的顾客们。这样，自由百货于1875年在伦敦开业，店内销售从东方异域运来的家具和各类器物，同时还有"唯美"服装，灵感源自古代的宽松长裙，这在盛行的紧身束身衣之外给人们提供了全新的选择。一些百货公司还在其他的市镇或城郊开设分店，其中就有1877年全英国首家专门设计建造的百货公司，即位于伦敦南部布里克斯顿的乐蓬马歇百货。还有一些百货公司在时髦的海边度假区开设分店，比如马歇尔和斯奈尔格罗夫百货的斯卡伯勒分店，这家店只在假日季营业。百货公司的扩张把时尚商品带给更广泛的人群。大多数百货商店都拥有自己的制衣部门，在19世纪后半叶成衣普及后，它们也开始大批销售成衣系列。百货商店全力构建着与顾客之间的关系，通过服务、品质和价格赢得顾客的忠诚。

这些发展不但改变了人们购买面料和服饰的方式，同时还塑造了大众对于应该如何举手投足、穿衣打扮的观念。店铺的广告向人们展示着可被接受的审美标准，并宣传着时尚个性的典范。这些都基于时装的不断传播以及人们融入消费者社会的渴望,而消费者社会在世纪之初便已成形。尽管百货公司代表了资产阶级的理想,它们却向更广泛的人群敞开着怀抱。1912年塞尔福里奇百货在伦敦成立,它沿用了美国的方式,搭配公司标志性的绿色地毯、纸笔、货车,还设置了一个极受欢迎的"特价地下区"。百货公司开放式的设计让更多群体的人们可以走进其中,自由浏览。虽然奢华的店铺也许吓退了一些购物者,但仍有人愿意努力为这些店中的奢侈品攒钱,店中顾客的尊贵身份和时髦风格让他们无限向往。到了1850年代,公共交通方式的发展使得乘坐公共汽车和火车出行购物更加简单和便宜。大型城市中的地铁交通让这一过程变得更加轻松,也鼓励人们把"购物一日游"当成一种愉快而又便捷的放松与娱乐方式。

百货商店极力吸引着顾客,它们综合运用盛大的时装秀以及令人兴奋的新科技,前者把法国时装的魅力展现在更多受众面前。1898年,伦敦哈罗德百货安装了第一批在楼层间运送顾客的电动扶梯,引来大批民众并被媒体广泛报道。20世纪早期,美国商店纷纷举办系列法国时装秀,真实的模特们在错综复杂的伸展台上穿行,在精心布置的灯光下散发着光彩。这些华丽表演的名字让人想到它们那颓废与奢华的氛围。1908年,费城的沃纳梅克百货举办了一场以拿破仑为主题的"巴黎派对"(Fête de Paris),呈现了法国宫廷的生动场面。同时,1911年纽约的金贝尔斯则办了一场"蒙特卡洛"活动。店内的剧场建造起地中海风格的花园,安置了轮盘赌桌和其他道具,让成千上万的到访者感受到了真实的里维埃拉奢华风情。

百货公司把时装带给普罗大众，从布拉格到斯德哥尔摩，从芝加哥到纽卡斯尔，不断在时尚购物地区开业，尽管如此它们依然远远不是时装唯一的来源。精英们依旧定期光顾那些为自己家族世代服务的皇室制衣商和定制裁缝。微型的专业商业中心繁荣如常，且常顺着新的潮流不断涌现。譬如20世纪早期人们对镶着羽毛的大帽子的狂热，促使商店开始销售鸵鸟羽毛等镶边装饰。变化的风格和流行的饰物同样吸引着男性顾客。除了向富有男性销售珠宝和饰品的奢侈品店，还出现了面向年轻男性的商品，这一群体迫不及待地想要花掉自新兴的白领工作中赚得的金钱。和女性时装类似，舞台上的名流们也推动着男性时装风格的传播，此外还有越来越多的银幕明星以及运动明星。领带、领结、领扣、袖扣众多颜色和图案的不断变化，为每一季的男士套装带来了活力。

　　邮寄购物则是另一项重大发明，特别是在美国、澳大利亚和阿根廷这样国内城市间距较远、亲自前往购物比较困难的国家。百货公司都有各自的邮购销售部门，这主要得益于邮寄包裹业务的改进和有线电话的采用。芝加哥的马歇尔沃德百货拥有可谓业内最知名的邮购业务，它的产品目录用不断丰富的面向所有家庭成员的成衣时装吸引着美国人。运输方式的改进同样推动了这一行业，货物先后被运货车、马车运送，再到后来坐上火车。

　　因而到了20世纪的前几十年，消费主义已经发展到覆盖了不同性别、年龄和阶层的广泛人群。随着1920年代大规模生产的方式得到改良，市场上销售的时装和饰物品类增长得也更加迅速，面对日益激烈的竞争，商店必须努力高效地销售这些商品。已经立足的百货商店和专卖店此时又加入席卷西方国家的"联营店"("multiples"),它们可以被看作连锁店的一种早期形态。美国的连锁分店开始销售借鉴好莱坞影星服饰的价格低廉的时装,在全国广受欢迎。在英国,赫普沃斯公司从1864年建立时的裁缝铺逐步发展到在全英拥有男装门店的规模,并且仍旧在持续经营,如今已演变成了囊括男装、女装和童装的连锁商店奈克斯特。连锁商店具有集中采购和制度化管理两个优势,可以确保价格实惠的同时管理市场营销和广告活动。它们力图打造一种统一的格调,包括店面设计、橱窗以及店员制服。20世纪后半叶,连锁店的统治让人们对同质化大为诟病,而且讽刺的是消费者真正的选择变少了,但是熟悉的各个品牌在各自的分店里为消费者稳定供应着款式、质量相同的货品,这给消费者提供了保障。

与之相反的是,高级时装设计师将沿袭几个世纪的传统和当代创新方式结合起来,继续销售着他们的设计作品。客人们享受着一对一的服务和个人定制,高级时装沙龙还合并了精品店以出售成衣线早期风格的产品,还有为取悦自家这些精英顾客而特别设计的香水以及各种奢侈品。不管是只为私人顾客开放的高级时装沙龙,还是时装秀期间的精选店买手及他们的精品店,都运用了现代设计和陈设技术来彰显它们的时装潮流。1923年,玛德琳·维奥内重新改造了她的时装店,运用了流畅的现代主义线条和古典风格的壁画。自1930年代中期开始,艾尔莎·夏帕瑞丽着手打造了一系列超现实主义的橱窗陈列,有力宣传了她设计作品中的才思。上面两个例子中,这些艺术借鉴都与她们的服装设计理念息息相关,也跟她们的品牌精神、整体包装和广告宣传遥相呼应，为消费者创造出能够一目了然的连贯的品牌风格。

高级时装设计师们需要投射出一种专属形象，为每一件挂上他们名号的产品都笼上一层奢华的光环。尽管时装已经逐渐成为全球成衣销售行业的一件工具，许多商店依旧认为巴黎是时装新风格的重要源头。例如，美国百货公司和时装店每季都会派出买手前往法国首都采购一批"款式"，他们会取得相关许可，对这些服装进行限定数目的复制并在自家店内销售。这些设计在店铺在售的系列中将拥有最高时尚地位，此外还有大致参考巴黎流行趋势设计的其他服装，以及不断增加的化用法国风格的本土设计师作品。买手们因而扮演了一个至关重要的角色，因为他们需要理解自己所代表店铺的时尚形象和顾客的喜好。保持店内货品不断的新鲜感对于店铺可谓至关重要。1938年，梅西百货副总裁肯尼斯·柯林斯写信给致力于促进美国时装发展的时装集团（the Fashion Group），信中说道：

　　……零售业有一条至理名言：时装商业的成败取决于商人迅速加入最新时尚潮流以及在潮流衰退时同样迅速脱身的能力。

这种新潮款式的快速更新是时装产业的立足之本。

萨克斯第五大道精品百货这样的大型百货公司都会开辟几条针对不同消费者的产品线。从1930年开始，公司销售起老板夫人苏菲·金贝尔设计的独家奢侈品，并以"现代沙龙"为品牌名称，此外还有她为公司专门从巴黎挑选的时装。后来又有了各种类型的成衣线,包括运动服饰以及以年轻女大学生为目标人群设计的服装,也有类似的男性时装产品。所有这些系列综合起来为萨克斯赢得了时尚口碑,充分展示了公司在为所有客户群体提供美好衣饰上所体现出的审美和眼力。这些系列在店中专门设计的区域里销售,以反映自己的受众及目标,公司还会在每年的几个关键节点在时尚杂志和报纸上投放广告,促使销量最大化。

大萧条时期,许多店铺无奈之下暂停派人前往巴黎,并越来越依赖本土发展起来的时装。尽管经济低迷,《时尚》和《哈珀芭莎》这些时尚杂志依然为各种规模的店铺投放广告。不同国家版本的《时尚》中类似"店铺猎手"这种专栏不断怂恿着女性出门购物,并规划好"最佳"的购物区域还有最时髦的精品店和百货公司。设计师和商店通过各自的媒体代理人与时尚媒体构建起密切联系,这些代理人努力争取杂志里的广告版面和专题报道。这种关系在接下来的几十年中不断延续。不过,第二次世界大战以及随之而来的物资短缺中断了商品的流通和供应。尽管大多数被卷入战争的国家都面临着供应短缺因而只能定量配给,许多国家仍然坚持把消费用品的梦想作为激励人心的远景。

随着经济在1950年代恢复,新的举措开始发展起来。其中一个关键例证就是伦敦的设计师自营精品店在这个十年接近尾声时大量增长。这些店充分说明,只要充分理解受众,知道什么样的衣服是他们想穿的,就算是小规模的从业者也能推出时装。比如,玛丽·奎恩特对当代时装现状的失望促使她于1955年在伦敦国王大道开出了自己的"集市",她对当代时装现状有着自己的困惑：

> 一直以来我都希望年轻人能拥有属于他们自己的时装……彻头彻尾的二十世纪时装……但我对时装商业一窍不通。我从不把自己看作设计师。我只知道我一心想要为年轻人找到合适的衣服，以及与之搭配的合适的配饰。

奎恩特女士制作出许多有趣的服装：娃娃裙、灯芯绒灯笼裤，还有水果印花围裙装，这些服饰帮助塑造了这一时期的时装风格。她与同时期的设计师们培育出的模仿者遍布全球，渴望能够利用这股年轻人驱动的大规模生产服装的潮流。她也给未来的设计师-零售商们提供了一个发展样本，他们将来会通过为新兴青年文化设计服装而在全球树立声望。薇薇恩·韦斯特伍德和马尔科姆·麦克拉伦的店铺也开在国王大道上，它们变换着店铺的内外设计以及在售服装的款式，与不断演变的街头潮流保持一致。1970年代早期店铺为售卖以"不良少年"为灵感的西装的"尽情摇滚"（Let It Rock），到1970年代中期变身为崇尚硬核朋克美学的"煽动分子"（Seditionaries）与"性"（Sex），最后化身为"世界尽头"（World's End）——一间"爱丽丝梦游仙境"风格的精品店，地板肆意倾斜，时钟指针倒走。韦斯特伍德的设计和零售环境风格皆是不断流动变化的亚文化的一部分。不同的时装风格随着音乐、街头文化，还有与时装相关的艺术界的前进而涌现、变化。这种灵活性为她的店创造出一种令人兴奋的群体感与流行感，汲取自各种亚文化中的"自己动手"精神进一步强化了这一气质。正如1960年代奎恩特女士的店面,它证明了精神气质一致的店铺可以聚在一起促成生意,同时也能巩固这一区域的时尚口碑。21世纪早期,纽约的字母城也出现了设计师制衣商在同一区域密集开店、荟萃一堂的盛况。

确实,西班牙连锁品牌飒拉以糅合经典款和大牌走秀款而闻名,其商业成功建立在品牌对购物区这种有机发展的战略洞察的基础上。自1975年开出首家门店起,飒拉已经扩张到全球,在高街时尚的竞争中压倒了其主要对手。每家飒拉门店都设计得像大牌精品店,主题鲜明的服装和配饰搭配成组,为消费者

图14 飒拉门店与大牌精品店外观相似,运用开放的临街位置与精心调配的展示吸引消费者走进店中

展示各种穿搭建议。连锁店隶属于印第纺集团,集团旗下业务还包括马西莫·杜蒂(Massimo Dutti)、巴适卡(Bershka)和飒拉家居。它采取的策略通常是先开一家大型飒拉门店并按照旗舰店模式运营,从视觉上呈现品牌精神,而后相邻开出集团其他分支品牌的店铺。这样的安排鼓励购物者在不同店铺间来回走动,购买印第纺旗下的不同品牌,同时又能让顾客亲自感受到每一家的服装、配饰和家居软装是如何相互搭配补充的。与此相联系的是飒拉对时装流行趋势的快速反应,它有一支小型的设计团队和一个紧密的生产体系,这使得新款式被发掘后能够快速转化生产出新的服装,并且在人们了解这一新潮流后不久便迅速送达各个门店。

其他国际品牌则依靠自有设计团队的能力来打造平价时装,并配以名流和最新款式系列。海恩斯莫里斯推出了许多联名系列,有些来自维克多与罗尔夫、斯特拉·麦卡特尼、卡尔·拉格斐这样的设计师,也有的出自麦当娜和凯莉·米洛这样的流行音乐明星。这些合作时装系列通常只持续非常短暂的一个时期,引来大量媒体竞相报道,并在系列发售时吸引大批购物者排队等待购买。这一方法的成功类似于20世纪高级时装设计师的成衣线与授权生产。高级时装的光环被用来提升各类面向大众店铺的时尚感,不管是美国连锁商店塔吉特(Target)还是英国时装零售商纽洛克(New Look)。这类合作里面最著名的大概是超模凯特·莫斯与Topshop的联名,这家英国的连锁店从1990年代后期就开始引领着高街时尚的发展。这是一次有趣的开发,它将明星的个性特质、时装风格和独有的光环转化为品牌遍布全球各地分店里的常规产品系列。系列中的服装仿效了莫斯私人衣橱中的单品,既有古董服装又有设计师作品。莫斯自己就堪比一个品牌,用来营销这些产品,甚至还启发了新的装饰元素,其中就包括她背上那对燕子文身,这一图样如今已然装点在牛仔裤、衬衣等各种服装上。名人与时装之间的联系起码自18世纪以来就十分明显,如今这种联名合作更是前所未有地深化了这一联系。

这一合作充分展示了20世纪晚期以来奢华服饰与大众时装之间愈来愈模糊的界限。在英国,凯特·莫斯联名系列在Topshop自营的高街店铺进行销售,人们因此将其视为某种由时尚引导却毫无疑问是大批量生产的抛弃型时装世界。纽约则不太一样,这些系列只在独家精品时装专卖店巴尼斯发布,被赋予一种专属奢华品牌的气质,与全球各地知名高级时装设计师品牌一起售卖。

这种高级时装与大众时装的混淆,实际上是过去150年来成衣时装不断发展壮大以及高街时尚产品强大的时尚化设计理念共同作用的结果。当消费者们更加自如地将古董服饰、设计师时装、廉价高街服装还有自己在市场上淘来的单品混搭在一起时,这些服饰品类的分界在某种程度上已然消解。虽然价格依旧是最显而易见的差异,但差异更多地体现在消费者将它们搭配出有趣而充满个性造型的能力,而不是坚守哪种衣服更加体面的固有观念。这一变化不单单体现在高街时尚的发展上。1980年代起,奢侈品牌开始扩张自己的领域,从只为精英服务的小型精品店发展到建立在大城市里的巨型旗舰店,同时也开始在免税店和专门销售过季超低价时装的购物中心里售卖自己的产品。

20世纪晚期，像古驰（Gucci）这样的奢侈品牌已经发展成为庞大的企业集团，并很快将远东地区认定为自己产品的重要市场。商店不仅开设在日本与韩国主岛，还同样开设在时尚人群的度假胜地。包括夏威夷在内的旅游目的地的酒店汇集了各家奢侈精品店，以供年轻富有的日本女性购物。综合的广告大片中，博柏利和路易威登这类老牌特有的经典传承与克里斯托弗·贝利、马克·雅各布这些新任年轻设计师为品牌强化的前卫时尚形象得到了恰到好处的平衡。包括高端电商网站颇特女士（net-a-porter.com）在内的线上时装商店使购买这些大牌时装更加简便。许多网站采用了杂志式的排版，提供独家商品、时尚新闻与风格参考，展示最新时装系列的大片和视频，并就如何打造全身造型提出建议，所有显示单品都附有购买链接。

到了21世纪之初，东方已经成为大众服饰和奢侈品时装的中心。这里既制造着供应本地市场的服装产品，也供应着全球大部分其他地区的产品线，日益富裕起来的市民对购买时装也热切起来。先后担任古驰和圣罗兰创意总监的汤姆·福特，认为这标志着时装的国际平衡出现了根本性变革。达娜·托马斯在《奢华：奢侈品为何黯然失色》（*Deluxe: How Luxury Lost its Lustre*）一书中引述了福特的评论：

> 本世纪属于新兴市场……我们（的事业）在西方已经到头了——我们的时代来过了也结束了。现在一切都关乎中国、印度和俄罗斯。这将是那些历史上崇尚奢华却很久没有过这种体验的众多文化重新觉醒的开始。

然而,时装产业各个领域的全球化进程引发了一些道德议题:一方面,远离公司管理中心进行生产制造活动可能存在劳动力压榨问题;另一方面,大品牌统治世界绝大多数市场后,消费社会带来的同质化结果也引发了人们的担忧。

第五章
道　德

　　1980年成立于美国的善待动物组织（People for the Ethical Treatment of Animals, PETA）如今已成长为保护动物权益的全球压力集团。它发起的运动涵盖了许多与时装相关的议题，强迫人们正视动物产品的使用，例如皮草和羊毛制品。2007年的一张宣传照展示了英国流行歌手、模特苏菲·埃利斯·贝克斯特身着一袭优雅黑色晚礼服的形象。她的面庞妆容精致：鲜红的双唇，白皙的皮肤，化着烟熏浓妆的眼睛。

　　再看下去，这个"蛇蝎美人"的造型就名副其实了：她单手拎着一只惨死狐狸的尸体，它的皮毛已被剥掉，露出猩红的血肉，头颅怪诞地耷拉在身体一侧。下方的标语写道："这就是你那件皮草大衣剩余的部分"，进一步强化了支撑皮草生意的残酷这一信息。这张宣传照的整体美学风格基于怀旧的黑色电影影像，而1940年代的电影女主人公们却经常肩头围裹着狐裘披肩，以作为奢华与性感的象征。PETA颠覆了观看者的期待，让人们不得不直面皮草背后的杀戮和恐怖。

图15　PETA使用醒目的图像与富有冲击力的方式揭露皮草交易的残酷

其他的印刷品和广告牌宣传照也运用了名人面孔与熟悉图像的类似组合,以及揭露时装产业阴暗面的这种并置所带来的冲击。这一组织的目标就是要迫使消费者认清时装大片和营销手段那愉悦感官的表象背后究竟发生着什么,并让人们学会审视服装生产的方式和相关过程。PETA的标语运用极富冲击力的直白广告语言组织出令人记忆深刻的口号,并进入人们的日常口语。典型的例子包括揭露皮草交易核心矛盾的那几句讽刺的双关语:"皮草是给动物准备的(Fur is for Animals)","裸露肌肤,而不是穿着熊皮(Bare Skin, not Bear Skin)",还有提倡将文身作为另一种时尚身份象征的"用墨水别用貂皮(Ink not Mink)"。

PETA对动物皮毛这一主题的关注意味着这些作为皮草来源的鲜活生命间的关联将被持续不断地再现。1990年代中叶,它曾做过一场著名的"宁可裸露,不穿皮草(I'd Rather Go Naked Than Wear Fur)"的宣传运动,请来一众超模和名流,她们褪去华服,立于巧妙安排的标语牌后。这些图片按照时装大片的风格设计制作。虽然身上一丝不挂,画中人仍旧被精心打扮和打光,以突出她们的"自然"美。通过模特、演员和歌手们的"本色"出演,她们自身的文化地位和价值观与PETA的宣传号召力之间形成了一种直接的连接。它传递出这样一种信息:假如这些广受欢迎的专业人士都拒绝皮草,那么普通消费者更应该如此。当然也有一些失足之人,比如娜奥米·坎贝尔,她在1990年代末脱离PETA,很快就喜欢上了皮草制品还有狩猎,但这些人也没能削弱PETA传递信息的强大影响力。21世纪初,包括演员伊娃·门德斯在内的一批新人与PETA签约。制作的图像包括名人裸身怀抱兔子的"放过兔子(Hands off the Buns)"宣传图。

PETA提升了时装产业对动物权利的认知。对于那些被该组织认定对时装中皮草的持续使用负有责任的对象,PETA成员闯过秀场,洒过颜料,还扔过轰动一时的冰冻动物尸体,此外也推动制定了针对羊毛贸易中对待山羊的新法规。PETA活动者的工作不仅强调了皮草贸易中不必要的残酷,同时还阐明了皮草如何常被错误地当成一种"天然"产品供人穿着,而事实是绝大多数皮草来自人工养殖,一经从动物身上获取,还需要经历各种化学处理以去除上面粘连的血肉,预制成待用的面料。

尽管PETA的目标令人钦佩,但他们所采取的方式引发了更多的道德问题。这一组织对时装视觉语言以及更广泛的青年文化的借用,致使人们指责它打着动物权利的旗号而持续对女性性化剥削。举一个著名的例子:英国"尊重"组织(British-based Respect)1980年代的宣传图"一顶皮帽子,两个坏了的婊子(One Fur Hat, Two Spoilt Bitches[①])",画面展示了一名模特身披一件死去动物制成的披肩,人们认为这一宣传照全然将女性当成了愚蠢、性感的物件。这种对立给宣传想要传递的信息带来了问题。有人认为这也可以被解读为博人眼球的另一种途径,让人直面穿着皮草这一行为的疏忽,通过令人瞠目的方式使人警醒。然而,为了做到这一点,它采用了主导当代多数广告的极富性意味的视觉方式。这一争议突显了这些宣传中自相

[①] "Spoilt Bitches"为双关,"spoilt"有"损坏"和"宠坏"两重含义,"bitches"也有"母兽"和"婊子"两层意思。——编注

矛盾的推动力。在大量的注意力关注一个伦理问题的同时，另一个同样重要的道德议题也被人们接受并且充满争议地信奉为现状。

时装产业的地位是模棱两可的。它既是一个利润丰厚的跨国产业，是许多人汲取愉悦的源泉，但同时又引发了一系列的道德争议。从对女性形象的描绘，到服装工人被对待的方式，时装兼具代表其当代文化之精华与糟粕的能力。因此，一方面时装可以塑造与表达另类以及主流身份，另一方面它又是如此专制、冷酷无情。时装所热衷的组合与夸大有时会破坏和混淆，或者甚至强化负面行为和刻板印象。时装对于外表的关注导致它常常背上肤浅与自恋的骂名。

坐落于洛杉矶的T恤生产商AA美国服饰（American Apparel）是另外一个典型的例子。公司宗旨自1997年成立之初便是致力于摆脱生产外包，试图打造"非血汗工厂"生产线。与其他专注基本必备款的品牌不同，这一品牌拒绝在发展中国家生产服饰，因为这些地方很难保证对工人权利与工厂环境的管控。相反，AA美国服饰选择用本地工人，通过这样的方式回馈当地社区。品牌店铺请来本土、国内知名的摄影师进行店内展览，店里那些酷酷的和都市感十足的基本款在各国市场大受欢迎。品牌的广告宣传强化了其道德认证，关注工人群体，常常请自家店铺中的店员及管理职员来做广告模特。

但品牌运用的图片风格却再一次引发了广泛的批评。AA美国服饰的老板多夫·查尼喜欢一种类似快照式的摄影风格——少男少女的偷拍照，他们常常半裸，对着镜头扭曲着身体。杰米·沃尔夫在给《纽约时报》撰写的一篇文章中曾写道：

> 这些广告同样充满了强烈的暗示意味，不仅仅因为它们在画面中展示了内衣或是紧紧包裹身体的针织衫。画面中的这些少男少女躺在床上或是正在淋浴；如果他们正悠闲地躺在沙发里，或是坐在地板上，那么他们的双腿则恰好是张开的；他们身上往往只穿着一件单品，否则便是一丝不挂；两三个少女看上去处于一种过度兴奋的欢愉状态。这些画面有一种闪光灯照射的低保真的闷热感；它们看着不像是广告，更像是某个人的聚友（Myspace）主页发布的照片。

这种美学并不新鲜；它借鉴了1970年代南·戈尔丁以及拉里·克拉克的青年文化图像。时装图片也不是头一回运用这种美学：卡尔文·克莱恩数十年来也组合运用了类似博人眼球的年轻模特大片来宣传自己的简洁设计。这一美学渗透影响了各类时尚杂志和线上社交网站，当然也包括AA美国服饰自己的网站，它在页面上将图片按系列划分展示，方便访问者浏览翻阅。因而在这些模特生活化的姿态以及随意的性感中，他们运用了一组熟悉的视觉符号。

AA美国服饰在视觉形象中运用的享乐、性感美学或许会让人以为这是一家面向年轻群体的公司，但它又与传统观念中关注道德议题的"令人起敬"的公司应该被呈现的方式格格不入。就像抵制皮草的广告那样，当一件产品或某个事业被置于道德层面上，对于潜在暧昧且性感的图片的使用尤其容易遭到公众的批判。如果当代价值观中的某一方面被人们提出，它也会提高人们对某一组织或品牌产出成果各个方面中潜在问题的意识。尽管AA美国服饰使用的图片与品牌所面向的年轻群体的口味相投，可与此同时它却运用了业余的情色美学，这种美学广泛影响了21世纪初期的文化。鉴于时装产业的道德地位如此令人担忧，而它在构建当代文化中扮演的角色亦受到重重质疑，人们能察觉到传播手段和表现方式会破坏道德信息和行为也就不足为奇了。

身份与反叛

与时装生产方式相关的道德议题自19世纪晚期以来引发了大量的关注，而推动早期批评的却是时装改变一个人外貌的种种方式。人们的道德担忧主要集中在时装的伎俩上，它既提升了穿着者的美貌或是身份，又混淆扰乱了社会礼仪以及可接受的穿衣打扮、行为举止的方式。时装与身体的密切关系以及服饰对身体缺陷的修饰功能，同时又为身形增添了感官诱惑力，这加剧了卫道士们的担忧：既担忧着装者的虚荣，又担忧观看者受时装的影响。历史上，相比想要赞美时装的文字，有更多文字认定时装意味着自恋、傲慢和愚蠢。比如14世纪的文字和绘画都认为过分注重外貌是罪大恶极，因为不管对男人还是女人而言，这种行为都标志着他们的头脑只在乎表面和物质而对宗教默祷不屑一顾。着装者们用时装创造出新的身份，或是颠覆应当如何着装的传统观念，这意味着时装挑战了社会和文化边界，并且让观者感到迷惑。这些焦虑始终占据着中心地位，在那里偏离规范极有可能会给时装和它的信奉者带来道德上的愤恨。

虽然人们希望体面的女士男士们在他们的着装中展示对当

下潮流的意识，但对细节过多的关注却仍然是个问题。同时，人们也认为时装对老年群体、底层阶级来说不合时宜。尽管如此，这些看法也没能阻止时装的传播。17世纪，本·琼森在他的喜剧《艾碧辛，或沉默的女人》(*Epicoene, or The Silent Woman*)里就作了评说，揭示了使得时装含糊不清的一些关键问题。剧作中，外表平平的女性被认为更加品行端正，而外表美丽的女子则被指责勾引男人。剧中同样谴责了试图追赶时装和美妆潮流的年长女性。剧中人物奥特评价自己的妻子：

> 那么一张丑恶的脸！她每年还花我40镑去买那些水银和猪骨头。她所有的牙都是在黑衣修士区做的，两根眉毛是在斯特兰德大街画的，还有她的发型是在银街弄的。这城里每一处都拥有她的一部分。

这种可以花钱买到美貌的观念，在这部剧里包括能够把脸变成时髦苍白色的水银，深深突显了时装与生俱来的表里不一。奥特夫人的购物之行意味着她的外貌更多地属于流行的零售商而非自然天成。因此，她不仅是在欺骗她的丈夫，还愚蠢地花着大把的钱想重回青春。

这一主题在接下来的各个时期里通过布道词、宣传册、专著以及绘画不断发展。18世纪晚期到19世纪初，讽刺画家，最著名的有克鲁克香克和罗兰森，画过一些装上假发、精心打扮了的年长女人，她们的身子用各种垫料和箍圈重新塑形，打造出身体曲线，使其符合当下的审美理念。1770年代，人们取笑最多的是高耸的假发顶上还插着一英尺长的羽毛；接下来的十年里，嘲笑声则转向裙身背后的臀垫；到了世纪之交，又开始嘲弄本就纤瘦的女子在身着最新潮的筒裙后看上去更加瘦弱，奚落丰腴的女士穿着同样的流行时装则看起来更加臃肿。

　　这种种批评反映了人们对女性、女性身体以及女性社会地位的态度。女性当然被视为不如男性重要，卫道士们还要监督着她们的衣着、仪态、礼仪还有举止。社会阶级同样扮演了重要角色，人们对精英阶层和非精英阶层的女性有着不同的标准与期望。有一点极为重要，所有女性都应当保持体面的形象，将自己与风尘女子清晰地区别开来，避免让自己的家人蒙羞。因此，女人们必须精心考虑如何使用时装；过多的兴趣容易引人非议，可兴味索然也会导致女性遭人质疑。时装在性别塑造中扮演的角色意味着它对于人们对自己的个人与集体身份的投射是一个关键元素。男人们因为时装选择而受指摘的情况要少得多，但他们依旧必须维持好与自己的阶层和地位相一致的形象。然而，那些过于热衷时装的年轻男子们确实也会遭到强烈的道德谴责。18世纪初，《旁观者》(*The Spectator*)杂志把热衷打扮的学生描述为"一无是处"，像女人一样，"只会以'衣'取人"。这或许是男士时装在色彩、装饰及款式上已然绚丽纷繁的最后一个时代了，因此想要反叛则需要更大的努力。正如《旁观者》所指出的那样，要做到这一点就是要挑战社会预期，甘冒被斥娘娘腔的风险。

　　这类男性多数都遭遇过性取向甚至是性别的怀疑。1760年代到1770年代，这批"纨绔子弟"（Macaronis[①]）与他们最直接的前辈"花花公子"（Fops）一般，招来讽刺画家和批评家们

[①] 原意为"通心粉"。——编注

图16 18世纪的"纨绔子弟"们因他们浮夸的衣着风格与忸怩的行为举止备受讥讽

的嘲笑。这些以意大利面命名的年轻男子们用色彩艳丽的衣着炫耀着自身与欧洲大陆之间的种种关联。他们的服饰夸张地演绎着当下的时装潮流,以超大号的假发为特色,有时还会扑上一些或红或蓝的粉,而不是常见的白色。他们身穿的外套剪裁极为贴身,形成曲线延伸到身后,且常常被描绘成一副姿势做作的样子。"纨绔子弟"们在许多方面都冒犯了男子气概的典范;他们被认为柔柔弱弱,既不爱国又自负虚荣。各种组织松散的过于时髦的年轻男子群体取而代之,每一个群体的打扮都极力夸耀着自己的与众不同,并反叛着社会观念。其中包括法国大革命时的"公子哥"(Incroyables),还有19世纪英国的"时髦人士"(Swells)和"花花公子"(Mashers)以及美国的"时髦男"(Dudes)。每一个群体都运用夸张的打扮、异域风情的时装以及对发型和配饰的格外用心来突显自我风格,并以此声明他们要挑战传统的男性典范,因而也是挑战现状。

自1841年开始,《笨拙》(*Punch*)杂志就喜欢在嘲讽时装中找乐子,另外也刊载一些以时尚的名义穿着衬裙、束腰和裙撑以将身体扭曲成精致形态的女人的图片。伴随这些讥讽评论的还有医生们的严正警告——穿着鲸骨束身衣将会危害女性的身体健康,但这些声音却几乎无法削弱这类服饰的受欢迎程度。性别依然是一个主要的问题。为了女性化的形象,女性必须穿着这些内衣,可她们又因为穿着这些束身衣被人们指责为不理智。这种双重束缚又扩展到那些被视为过分男性化的服装,哪怕它们比起高级时装更为实用。1880年代,当女性走上白领职位,她们身着的所谓定制服装,即基于男式西服改造但加上裙装的套装,被认为将女性转变成了男性。的确,在所有这些例子中,服装确实被看作穿着者性别、性向、阶层以及社会地位的一种标志,任何不明确都可能引起误解与指责。

这一点清晰地体现在长期以来人们认定女性不应穿裤子的

观念中,这一观念认为女人穿裤子是对性别角色的扰乱,暗示着女人一心要夺取男人的支配地位。种种担忧一直延续到20世纪。1942年,女演员阿莱缇在巴黎目睹的穿着裤装的女性人数令她大为惊骇。虽然正处战时的艰难日子,她仍旧认为不应为这种行为找任何理由,并且:

> 对于那些完全有办法买到靴子和大衣的女人来说,选择穿裤子简直不可原谅。她们无法给人留下任何好印象,而这种缺乏自尊的行为只是证明了她们糟糕的品味。

这个例子不仅揭示了缺失女性特质可能给人带来的恐慌,同时也强调了类似这样的道德指摘中的社会因素。身处某些特定职业之中的劳动阶级女性,包括采矿业和渔业,自19世纪以来就穿着裤子或马裤。然而,她们实际上都隐形了:字面意义上,绝大多数身处她们日常职业环境之外的人根本看不到她们;隐喻意义上,全然是因为中产阶级和精英阶层根本无视她们。

在对时装如何可能掩盖一个人的本来地位甚至炫耀其地位以作为对权威的挑战的种种道德担忧之中,阶级是一个恒久的主题。到了20世纪,人们素来对藐视中产阶级体面与礼仪理念的服装的质疑,此时又因众多故意挑衅的亚文化群体的崛起而加剧。在1940年代初的法国,"先锋派"(Zazous)群体中男男女女们细节丰富的西服套装、墨镜,以及美式发型与妆容引起了社会的惊慌。大众与媒体对他们穿着的愤怒集合了一系列常见的问题。他国的时装风格在本国人看来是不爱国的,尤其是当时正处在战时的大环境下,即便美国人身为盟军。他们夸张的服饰和妆容打破了基于阶级建立起来的优雅品味的观念，同时张扬着好莱坞彰显自我的浮夸风格。虽然他们的风格一直只限于一个数量很少的年轻人群体，但"先锋派"们对电影明星时装的效仿以及对爵士乐的热爱却是对法国文化的一种视觉和听觉上的对抗，而此时的法国正处于被纳粹占领的威胁之中。

接下来的数十年中，青年文化持续上演着对关于行为表现的社会规范的破坏。在英国，阶级在塑造亚文化的本质上扮演了至关重要的角色。1960年代，"摩登族"模仿中产阶级的体面打扮，穿上了整洁、合身的西服套装，而"光头党"（Skinheads）则基于工装强化了这一风格，显示出强烈的工人阶级身份。在这几个例子中，青年文化都是在群体成员对于新奇事物的探索，以及对某种音乐风格的热衷这两者合力推动下发展的。到了21世纪早期，工薪阶级青年以及无业青年中出现了一个影响更广的群体。Chavs[①]总是被人们指责毫无品味，原因就是他们对身上醒目的品牌会不自觉地炫耀，对中产阶级所秉持的时尚理念也毫不尊重。媒体报道暴露了根深蒂固的阶级偏见，这个词很快就与政府住宅群的青少年犯罪产生了关联。Chavs身上具有攻击性的运动装与工人阶级负面的刻板印象联系在一起，成为一种可以轻松识别的内城区不法分子的视觉体现。

媒体对青年风格的每一种新形象的指责，展现了破坏现状的反叛行为所能带来的巨大影响。在日本，东京的原宿地区从1980年代开始就一直是街头时尚的焦点，青年人群从这里衍生发展出了许多穿衣搭配的新方式。少女们颠覆了传统观念中的

[①] 意为"赶时髦的年轻人"。——编注

女性形象，创造出令人惊叹的新风格，从种种来源中自由地组合各种元素，其中包括高级时装、过去的亚文化、动漫以及电脑游戏。事实上，她们这种混合的时装风格反映了电脑虚拟角色中天马行空的个人形象设计，后者在远东地区极受欢迎。原宿的街头时尚公然反抗了父母期望中女孩应当展现出的端庄矜持的形象。流行歌手格温·史蒂芬妮打造了"原宿女孩"四人舞团，在她的音乐录影带、演唱会现场演出，给这些风格又添上一层争议。韩裔美国喜剧明星玛格丽特·曹曾经批评过史蒂芬妮对这种亚洲时装风格的不恰当使用，认为她对这些"原宿女孩"的利用极其无礼，她强调："日本校服有点像扮黑人时用的化妆品。"她的意思是这些舞者代表了一种种族身份的刻板印象，不过是用来为白种人的表演增加气氛罢了。史蒂芬妮的时装风格本身受到了日本街头时尚的影响，但她这些伴舞把这种风格做了进一步的演绎。她们的存在确实不仅体现了对服装借鉴国外风格的担忧，更重要的是究竟谁有权滥用这些时装风格，同时还有对将种族形象刻板化的道德担忧。

此种担忧的另一个不同表现是，人们对选择戴着希贾布头巾以作为宗教信仰与种族身份标志的年轻穆斯林女性的那种困惑又常常有些过激的反应。后"9·11"时代里人们对于伊斯兰教的恐惧，加上公众以及媒体认为这种外在区别属行为失范的观念，使得在某些法国学校女孩们被禁止戴希贾布头巾。这引发了人们的抗议，也更加坚定了部分穆斯林女性对于希贾布头巾重要性的信仰，它不仅是她们宗教信仰的象征，也是对西方理想中的女性特质以及当代时装裸露身体的一种质疑。

这一议题使得在一些极其特定的案例中，针对少数族裔群体衣着和外貌展示或处理方式的道德抗议更加尖锐。模特圈子对非白人女性的忽略是行业内的一个重要问题。尽管媒体一直在抗议,杂志有时也会做一些一次性的特辑,例如意大利版《时尚》2008年7月刊的所有专题大片全部使用黑人模特,但白人女性依旧统治着T台,时装摄影和广告也不例外。作为非裔英国人的一员,行业顶级的模特乔丹·邓恩曾说过:"伦敦并非一座完全由白人构成的城市,那么T台上为什么一定要都是白人呢?"时装业对多元性的这种固执的无视,恰好是广泛文化范畴内固有的种族主义的表征。真实的模特以及她们在时尚杂志中的形象展现,迫切需要时装产业改变原有的态度,真正意识到继续只关注白人模特是不被接受的。

规则与变革

伴随着对时尚图片中男性,尤其是女性的表现方式的抗议,各种试图控制或管理时装生产与消费的尝试也不断出现。文艺复兴时期,禁奢法令一直被强制实施,通过限制特定群体使用特制面料或装饰类型来保持阶级区分,或者是强行向民众灌输节俭观念。例如,意大利通过立法来管制人们在诸如婚礼等仪式上的着装,同时也限制了不同阶层女性领口所允许裸露肌肤的尺寸。这些法令定期在整个欧洲实行,但效果依然有限,毕竟其监管实在难以实现。正如凯瑟琳·克韦希·基勒比在提到表达了对服装过度展示的社会担忧的意大利法律时写道:"这些法律本质上就是在自欺欺人:通过法律禁止奢侈品刚好采用的一种形式只会催生新的形式以避免迫害。"由于时装始终处在变化之中,虽然早期的变化速度稍慢一些,立法速度难以跟上这些变化的节奏,而且如基勒比所言,这些穿衣人同样富有创意,通过改变服装的款式以规避法律,并创造出某种时装风格的全新形式。

禁奢法令在17世纪式微,虽然它们在二战时期有过复苏且成效甚著。早期对外来货物的进口禁令是出于经济与民族主义的原因,但二战的时间跨度与范围意味着许多类似的法令又因海空战争大肆展开之后严格的国际贸易限制而加重。物资短缺导致许多参战国实行定量配给。1941年,英国通过发行全年可兑换服装的服装券来管制服饰的生产及消费。每个人所分得的服装券数量在整个战期及战后期有所变化,但它们却对获取衣饰实行了非常严格的限制。英国、美国以及法国的法规还规定了服装生产中允许使用的面料用量,并大大削减了可以使用的装饰数量。获取时装的这一严苛的转变在英国公共事业计划的推动下得到了缓和,这一计划雇用了包括赫迪·雅曼在内的许多知名时装设计师,设计出既遵守法律规定又时髦有型的服装。然而,新衣服的匮乏意味着人们很难规避战时的限制法令,公众与媒体对过度铺张也持严厉态度,认为这种行为缺乏爱国精神,与全民抗战的努力背道而驰。

战后,苏联阵营国家得以延续对时装的这类限制,并试图给时装加上反社会主义的罪责,这在各国收效不一。东德的贾德·施蒂茨写道:

> 通过将女性作为消费者的权利与她们作为生产者的角色联系起来,并且将理性的"社会主义消费者习惯"作为一种重要的公民素质来宣传推广,官员们努力引导与控制着女性的消费欲望。

然而，根据职业特点设计制造的服装，包括围裙和工装等，吸引力非常有限，而且正如在其他社会主义国家中一样，包括捷克斯洛伐克，在更具功能性的时装风格之外还发展出一种国家许可的时装与时装意象的不稳定的结合体。这些想要革新时装并实现服装的道德形式的尝试顺从了19世纪服装改革家的倡导，比如古斯塔夫·耶格尔博士，他提倡无论男女都应当抵制时装的过度铺张，并使用天然纤维的服装，还有欧洲、斯堪的纳维亚半岛以及美国的女权主义者们，她们呼吁服装应具有更大的平等与理性。

这些旨在管制服装并创造出不伤害动物、人和环境的服装的推动力在20世纪晚期和21世纪初的形式开始逐渐进入主流观念，同时也融入商业时装之中。在嬉皮士以及与之相关的1960年代至1970年代寻求更为天然的时装以及关注道德议题的运动刺激之下，21世纪之交设计师与顶尖品牌们努力地调和着消费主义的发展与人们对更周到的时装设计与生产需求之间的矛盾。自20世纪早期开始，人们采取了一系列措施来管理工人的薪资与工作环境。这是由1911年纽约三角大楼制衣厂大火等事件推动的，其中有146个移民工人丧生。没人知道这间工厂中有多少转包工人，他们领着微薄的薪水，在拥挤逼仄的环境中工作，这意味着许多人根本无法从蹿出的大火中逃生。虽然类似的事故引发了针对血汗工厂的广泛抗议以及对最低薪资保障的呼吁，这样的现象直到今天仍旧未能彻底消除。随着大城市租金上涨，大批量生产向更偏远地区转移，最终迁到了南美洲和远东地区更为贫困的国家，那里的劳动力和房屋都更为廉价。所谓的"快时尚"，即品牌竭力供应着的时装秀上刚刚展示的最新时装,引发了激烈的竞争,这些品牌要不断以尽可能的低价在全年投放新的款式。

热门的高街品牌使用雇用童工生产的供应商这一行为一直以来遭到了各种控诉。2008年10月,英国广播公司(BBC)与《观察家报》(*The Observer*)做了一篇报道,指称廉价品牌普利马克(Primark)的三家供应商使用印度难民营中的斯里兰卡幼童,在极端恶劣的工作条件下为T恤衫缝制装饰物。意识到自己的处境后,普利马克立即解除了与这批供应商的合约,但报道揭示了当代时装产业中的一个核心问题。廉价服饰的便宜易得让人们对时装的获取更加民主化,可是又变相鼓励着消费者把服装当作短期消耗品随意抛弃,再加上生产廉价产品的激烈竞争,自然使得剥削成为潜在的后果。面向大众的时装连锁品牌都宣称,是巨大的销量使得它们的服装价格变得亲民。但是,这一模式中却存在着道德以及人力的成本,因为供应链正变得越来越分散,越来越难以追踪。记者丹·麦克杜格尔是这样说的:

> 英国现在有句话叫"急冲到底"(rush to the bottom),就是人们用来形容跨国零售商们雇用发展中国家承包商的行为,这些承包商通过偷工减料来为西方出资人压低补贴,提升利润。

普利马克并非面临非议的唯一连锁品牌;其他品牌,包括美国的盖璞,它们的供应商同样出现了许多问题。像英国"人树"(People Tree)这样的品牌因此力图避开上述商业模式,它们与自己的供应商建立起密切的联系,努力地创造可持续的生产模式，在那些生产其产品的国家里为当地社区造福。更大的品牌，比如 AA 美国服饰，也采取行动，通过使用本地工人来防止出现血汗工厂现象。这两个品牌还努力使用对环境影响较小的面料。牛仔与棉花生产中漂白与染色工艺的毒害，促使有机和非漂白产品在市场各个层面上涌现。与前几十年生产的早期产品不同的是，如今的生产商意识到，即便对于合乎道德的产品而言，消费者也会期待它们有时尚的设计价值。小众品牌像是鲁比伦敦（Ruby London）在其产品中加入时髦的有机棉紧身牛仔系列，瑞典品牌 Ekovarnhuset 除了自有产品线还出售其他生态时装品牌，创造出既时髦又环保的服装。甚至海恩斯莫里斯、纽洛克、玛莎（Marks and Spencer）这样的大品牌都引入了有机棉产品线。高级时装也开始涵盖越来越多的合乎道德的品牌。斯特拉·麦卡特尼拒绝使用动物皮草或皮革，丹麦品牌 Noir 的设计师们则把前卫的时装风格与严格的道德经营方针结合起来，方针包括支持生态友好面料的发展。

　　此外也有一些设计师提倡"减少购买"但投资使用周期更长且较为昂贵的单品的理念。这种"慢时尚"理念下的产品系列就有马丁·马吉拉纯手工制作的"手工"系列服装。《纽约时报》的记者阿曼德·利姆南德尔分解了这些奢侈品的相对成本后计算得出：以一件拉夫·西蒙为极简风格品牌吉尔·桑德（Jil Sander）设计的定制男士西装为例，它的定价为 6 000 美元，需要花费 22 小时制作完成，也就是说平均每小时单价为 272.73 美元。尽管这一算法并不能预算出每一次穿着的成本，它却倡导人们转变态度，拒绝快速变换的流行风格与按季购置最新时装的行为。然而并非所有人都负担得起这些必需的初始投资。

不过，慢时尚指出了人们试图让时装更符合道德准则所做的努力中的一个核心问题：消费本身就是问题症结所在。时装给环境造成的影响覆盖了一系列问题，从棉花等天然纤维种植过程中的生产方法与行为，到大众消费主义以及公众对新款时装的渴求。

日本连锁品牌无印良品（Muji）的再生纱线针织服装系列提供了一种解决方案；巴黎的马里籍设计师克叙里·比约特在设计中使用回收的旧毛衣则是另一种思路。这些服饰都依赖二手织物与服装，可以视为对20世纪末转向古着与跳蚤市场购置时装行为的一种配合。这些时装对环境的影响更小，并减少了生产过程，但它们不太可能完全取代现有的时装产业，尤其是考虑到它巨大的国际化范围还有与其生产与营销息息

图17 远东地区的许多市场出售奢侈品牌It手袋最新款仿制品，价位仅为真品市售价格的零头

相关的巨额资金。

  道德购物本身也存在着异化为一股时尚潮流的风险。随着全球经济在21世纪头十年里衰退,许多报道不断质疑着"衰退时尚"(recession chic)与"感觉良好消费主义"(feelgood consumerism)这类理念,它们建立在人们购买有机或符合道德准则生产的服装时内心的美德体验,即便他们的购买实际上并非必需。问题停留在消费者是否自愿拥有更少物质,并减少把购物当作某种获取休闲与愉悦的渠道的行为,以及道德化的品牌是否能坚持对应该购买何种产品的评判并维持自己的发展。

  遍布全球的仿制品市场兜售着最新款的It手袋复制品,表明了身份象征具有永恒的吸引力,以及时尚能够诱发人们对具有奢华与精英风格物质的渴求。随着时装覆盖了所有社会层次,并吸纳了国际知名品牌,监管其生产或管制其消费都变得越来越困难。想要实现这一目标,只有靠大规模重组社会与文化价值观,并变革全球化产业模式。这一产业在数世纪以来不断成长,引诱消费者并满足他们对服装触感与视觉魅力的欲望。

第六章

全球化

曼尼什·阿若拉（Manish Arora）2008—2009秋冬系列以艺术家苏伯德·古普塔整齐布置的不锈钢厨具装置艺术为秀场背景。这一金属布景为那些关于印度文化的陈词滥调作了一次讽刺的注解。古普塔闪闪发亮的陈设同时也预示了阿若拉时装秀中主导的冰冷的金银色调。他的模特被装扮成未来女战士的模样，并混合运用了许多历史元素，创造出金光闪闪的胸甲、硬挺的超短裙，还有接合的下装。罗马角斗士、中世纪骑士以及日本武士形象都得到呈现，并通过带刺的面具来强化充满力量的形象。这些来自各国的灵感元素，在阿若拉标志性的色彩明艳的三维刺绣、珠饰及贴花工艺之下，被发挥得淋漓尽致。这些工艺进一步体现了其融汇古今的手法，它们展示了传统的印度工艺，使用光彩夺目的施华洛世奇水晶来增强效果。

阿若拉的合作者们也一样丰富多元。日本艺术家田明网敬一使用大眼娃娃和奇异野兽等迷幻形象，以作为裙装与外套的装饰图案。华特·迪士尼的高飞狗、米奇和米妮也戴上了护甲

图18 曼尼什·阿若拉在其2008—2009秋冬系列中融入了女战士形象与华特·迪士尼动画角色的刺绣图案

和头盔,在一系列服装上全新亮相。整个系列突显了阿若拉的能力,他能够自看似不相关的元素与想法中打造出一套完整的造型,同时又强化了他作为国际时装设计师的地位,能够通过他精美的时装设计消除东西方鲜明的界定。自1997年创立自有品牌以来,阿若拉已经创作了许多充满想象力的作品,融合了传统刺绣与其他各种装饰工艺,运用波普艺术风格的色彩搭配和数不胜数的借鉴元素。他的装饰显示出奢华与繁复的特点,同时又细致入微地记录了他在时装产业中的个人成长。在伦敦时装周办秀时,英国国会大厦与皇室阅兵庆典的全景照被密集地印制在伞裙上;之后在巴黎,裙子上出现的则是埃菲尔铁塔。从一开始,他的目标就是要打造一个全球化的奢侈品牌,同时迎合印度以及各国消费者的品味。确实,他的时装风格使得这些区别越发不合时宜。大多数情况之下,这些消费者之间没有任何差异,如莉萨·阿姆斯特朗所言,阿若拉"看起来并没有迎合国外市场——也没有试图弱化自己的繁复风格"。

21世纪早期见证了时装周在全球各地稳步增长的日程安排,时尚潮流通过互联网的即时传播,以及印度、中国等国财富与工业生产的增长。阿若拉的个人成功是印度作为时尚中心的自信不断发展的产物。印度的纺织品与手工技艺自古就声名远扬,但直到1980年代晚期才开始建设发展时装产业必备的基础设施。高级时装设计师开始出现,包括阿若拉曾就读的新德里国家时装技术学院在内的专业院校培养出一批新兴设计师。1998年,印度时装设计理事会成立,旨在推广印度设计师以及寻求资金支持。这使得成衣品牌有了发展的可能,也为在印度之外发展影响范围更广的时装产业打下了基础。阿若拉的商业能力使他获得了世界范围内的知名度，并为他带来利润丰厚的设计合作机会。例如，他为锐步（Reebok）制作了一个鞋履系列，为斯沃琪（Swatch）打造了一个限量腕表产品系列，还给魅可（MAC）设计了一个彩妆系列，展现了他标志性的明亮色彩和他对闪亮表面的热爱。像这样的商业合作为阿若拉提供了扩张自有品牌的平台。

尽管如此，他的成功不应当只以他在西方世界内的认可度来判定。相反，作为不断壮大的能够熟练操作国际化销售并获得关注的非西方设计师群体中的一员，阿若拉身上体现了时装产业核心的地位正从西方逐渐偏离这一趋势。这一过程绝没有终结；值得注意的是，尽管阿若拉在伦敦和巴黎的时装秀提升了其在国际媒体与买手群体中的形象，他仍然会在印度办秀。而印度中上层阶级的崛起意味着他和他的同侪们拥有巨大的潜在国内市场，这种情况同样出现在其他致力发展时装产业的国家，其中就包括中国。

西方时尚都市也从吸引国际设计师加入本地项目带来的声望中受益。伦敦时装周一直争取国外媒体与首要店铺买手出席时装周大秀，努力维持着自己的业界形象。2005年2月，记者卡罗琳·阿索姆和艾伦·汉密尔顿描述了阿若拉、日本的丹麦–南斯拉夫裔与中国裔双人设计师组合阿加诺维奇与杨等名字如何给时装周日程增添了趣味和多样性。这些国际设计师与伦敦本地的尼日利亚裔设计师杜罗·奥罗伍、塞尔维亚裔设计师洛克山达·埃琳西克，还有来自新加坡的设计师安德鲁·鄞同台展示。这些来自全球的名字汇聚一城，突出了时装产业的国际化视野，同时也表明，尽管民族风格与地方风格在过去或许有助于将设计师作为群体来推广,但当越来越多的时尚都市不断涌现,设计师们也在资金的支持下能够在任何地方展示自己的作品时,这些风格的区别便不再那么重要了。时装产业的地理已经发生改变,然而正如须摩提·纳格拉斯所言,"由于印度时装产业(举例而言)是全球时装界一个相对较新的成员,这意味着为了参与其中,'本土'产业必须努力在一个既有体系内运作"。然而,随着其他地区的发展,加上商品和劳动力流动改变了生产模式,19世纪末形成的时装产业的基础格局自身或许也开始转移重心。

如今,巴黎巩固了其在西方时装业的中心地位,不过,甚至在20世纪之初,法国时装业就开始对美国优异的经营方式倍感担忧了。一旦美国成衣在第二次世界大战时发展出自己独有的特色,不只是高定时装,成衣也有可能创造时尚潮流。随着时装在战后复兴后使用起美国模特,牛仔服和运动装等休闲风又赢得了国际市场的认可,时装业迎来了一次根本性的变革,尽管此时的巴黎依旧发挥着巨大的行业影响力。大概在21世纪之初,一次相似的变化进程又蓄势待发,而这并不必然是一次全新的发展。事实上,至少对于印度和中国而言,它代表了奢侈品和视觉夸示在这些国家的复兴,它们丰富技艺的悠久历史曾因殖民主义、动荡政局与战争炮火而中断。

贸易与流通

贸易线路自公元前1世纪起就将纺织品输送到世界各地,将远东、中东地区与纺织品商贸繁盛的欧洲城市连接了起来。意大利曾是东西方世界之间的一道大门,它将自己打造成奢侈纺织品贸易中心。北欧形成了羊毛制品中心，意大利则以其样式色彩丰富的昂贵丝绸、天鹅绒与织锦而闻名天下。威尼斯和佛罗伦萨等城市出产了欧洲的绝大部分精美织物，这些面料有时也会留下创造它们的地中海贸易活动的印记：伊斯兰、希伯来和东方的文字及纹样与西方的图案融合在一起。这些跨越不同

图19　文艺复兴时期的织物常常糅合了欧洲、中东和远东地区的各种图案

文化的元素借鉴是贸易活动的一种自然产物,随着各国努力控制特定区域或是探索新大陆,这些商贸活动在文艺复兴时期发展了起来。15世纪到16世纪,贸易活动在更多欧洲国家间不断壮大,打通了葡萄牙、叙利亚、土耳其之间,印度和东南亚之间,还有西班牙与美洲之间的线路。

17世纪早期,英国与荷兰先后建立起东印度公司,正式组织起它们与印度和远东地区的贸易活动。最初,如约翰·斯戴尔斯所说,英国东印度公司最感兴趣的是把羊毛出口到亚洲,并且只买回极少量来自东方的顶级奢华织物,因为它们的样式在英国的吸引力非常有限。不过,到了17世纪后半叶,东印度公司给自己的印度代理人先是带去图样,后来又带去样品,鼓励当地生产出符合英国人心目中"异域风情"的产品纹样。这些产品大受欢迎,同时也意味着西方时装在其影响下使用了这些材料后,吸收了更多的东方产品。欧洲积累和发展出成熟的航海知识和运输方式来保障其贸易,并不断开发利用着亚洲工匠的创新、变通和技艺。他们生产出品类丰富的原料,并能对消费者的喜好迅速作出反应。这为跨文化交流提供了肥沃的土壤,生产出融合不同国家与民族元素的款式。尽管如此,西方的品味仍占据支配地位,影响着亚洲图案的使用方式。消费者被鼓励欣赏这些来自遥远国度的风格,这些风格已经经过了深谙其品味和欲望的东印度公司代理人的改造。驱动全球织物贸易的正是人们对奢华面料感官体验的渴望,还有西方世界对于新兴的异域风情的兴趣,其巨大的赢利潜力更是推动了这一活动。这一点建立在精英阶层对奢华展示的欲望之上,而这种欲望在所有国家都是共通的。

服饰风格总是趋于保持其独特性，然而有一些特定类型的服装却是从东方演变到西方来的，这里面就包括欧洲男士和女士在家中非正式场合所着的土耳其长袍式服装及围裹式长衣，以及17世纪末掀起的一股类似的穆斯林头巾风潮。这一时期的肖像画中，西方男性身着闪光绸质地的裹身外衣休息放松，精心修剪过的头顶上包裹着穆斯林头巾，以此作为在公众场合穿戴扑粉假发套之外的一种令人愉快的逃离。确实，彼得·斯塔利布拉斯与安·罗莎琳德·琼斯就曾有过论述，17世纪人们的身份与国家或大洲概念的联系不再那么紧密。他们分析了凡·戴克1622年所绘的英国驻波斯大使罗伯特·舍利像，以此证明精英的资格在这一时期是身份更为重要的组分。舍利身着与其社会阶层、职业身份相宜的波斯装束。他衣饰上华丽的刺绣、金色背景下色彩鲜艳的绸缎，充分展示了东方的这些技艺是如此纯熟，波斯的服装是如此奢华。斯塔利布拉斯与琼斯指出，舍利不会认为自己是个欧洲人，因为这一地区在当时尚未形成一致的身份认同。他也不会因为自己的西方人身份而产生优越感。他们认为，舍利会很自然地把波斯服饰作为自己新职位的一个标志，同时也将其作为对伊朗国王恭敬之心的一种表示。时髦身份同样与阶层和地位关联着，但与之相关的还有不同地区或宫廷的审美理念以及个人接受与诠释当下潮流的能力。然而舍利的肖像表明，这一身份在特定的社会或职业环境中可能会吸纳其他民族的期望这些元素，尤其是在国外生活或旅行的时候。在其后一个世纪里欧洲女性中流行的土耳其式宽松裹身裙进一步证明了这一点，它们实际上是像玛丽·沃特利-蒙塔古夫人这样的女性旅行者对真实的土耳其服饰进行的改良。

事实上，似乎17世纪时奢华与夸示的观念无论在东方还是西方世界的贵族和王室圈子里都是十分普遍的。卡洛·马可·贝尔凡蒂指出，17、18世纪时尚风潮在印度、中国和日本发展了起来，其中某些特定的审美与风格类型在当时大受欢迎。比如，在莫卧儿帝国时期的印度，服装制作中人们喜好繁缛的设计，头纱和头巾风格流行一时。衣饰剪裁和设计的风潮也开始在大城市的文职人员身上出现。不过，贝尔凡蒂认为，尽管时装自身在东西方世界中同步发展着，但它并未在东方成为一种社会制度，而到19世纪被禁止的服装形式成为一种社会常态。

不同文化间的借鉴却超越精英人群，它体现了基于贸易活动却有赖于吸引东西方消费者的设计的全球化影响。西方世界演化出自己对东方服饰设计的独特解读。18世纪中叶，中国风（*chinoiserie*）装饰潮流席卷欧陆。艾琳·里贝罗描写了这些对东方世界的再想象，它们创造出各种印满宝塔、风格化花卉，以及其他改良过的中式图案的纺织品。我们可以认为这类风潮部分来自贵族们对于衣装打扮的热爱，本例中表现为对其他民族文化风格的幻想形式的解读。中国成为化装舞会的热门主题，瑞典王室甚至在皇后岛夏宫给未来的国王古斯塔夫三世穿上了中式长袍。

中国风是西方对东方服饰奇幻想象的一股风尚。而18世纪时印度棉布空前的流行则表明，印度纺织品生产与印花设计对市场的影响能扩散到欧洲之外，一直延伸到各国在南美洲的众多殖民地等区域。大量印度棉布的低廉价格意味着其覆盖的人群范围前所未有。这也意味着纺织品设计与风格的全球化审美、大众获得时装的途径以及易于清洗的衣饰，对社会各群体（极度贫困人群除外）而言都触手可及。事实上，到了1780年代，所谓的"印花棉布热"引发了各国政府的恐慌，他们害怕各自的本土纺织品贸易会被淘汰。许多国家都通过了限制法案，包括瑞士和西班牙。玛尔塔·A.韦森特写道，据传在墨西哥，女人为了买这些外国时装竟然会出卖色相。然而，最终西方国家在这场传播迅猛的时装大潮中发现，比起与它的热度做斗争，他们更应该好好利用它来构建自己的纺织工业，并且运用从印度织物生产商身上学到的东西，在这场热潮中好好赚一笔，比如巴塞罗那就是这么做的。

这成为将要到来的一场重要全球转变的一部分——从不断创新又适应性强的印度纺织品贸易转向越来越趋于工业引领的西方世界，这一转变在19世纪加快了步伐。尤其当英国取得了一连串用于提升纺织品生产速度的发明成果之后，它取代了印度的纺织品生产，使得印度手工织造纺织品在1820年代几乎被彻底抛弃。随着西方国家开始更加依赖自己的面料生产与出口，而不再依靠进口棉布，时装业在纺织品生产领域的权力平衡发生了改变。西方时装体系迅速出现，其形式在未来一个世纪乃至更长时期内都占据着支配地位。机械化先后使得欧洲与美国的纺织工厂能够对变化的品味和时装潮流迅速做出反应。1850年代，欧洲发明的合成染料，尤其是威廉·珀金发现的颜色艳丽的苯胺紫染料，几乎彻底摧毁了世界其他地区的天然染料工业。桑德拉·尼森写道，这种染料使得这些新鲜而生动的色调风行全球，改变了从法国到危地马拉等各个角落的传统与流行服饰的面貌。

整个19世纪的时间里，西方国家对殖民地不断征服的过程见证了欧洲列强对纺织品贸易的剥削。尽管维多利亚文化中充斥着显著的种族主义态度，但精英阶层和中产阶级的消费者却仍旧钟情于来自欧洲之外的各种产品，其中包括印度的纺织品与日本的和服。亚瑟·莱森比·利伯蒂1875年在伦敦摄政街开办了他的百货公司，里面销售着来自东方的家具和装饰品，由于老板喜欢更加宽松、颜色更加柔和的亚洲风格和中世纪欧洲的垂褶长裙，店内也销售着以此为灵感设计的服装和纺织品。然而，佐藤知子与渡边俊夫证明利伯蒂对东方的态度是矛盾的，而且表达了西方世界对异域风情的幻想与亚洲真实面貌之间的棘手关系。1889年，利伯蒂在日本待了三个月，像其他同代评论家一样，他欣喜地发现在西方的影响下，丝绸变得更纤薄，也更易加工，但利伯蒂不喜欢东方丝绸在颜色与图案上的变化。从日本在1850年代向西方重开国门、开始现代化进程的那一刻起，无论男女都在传统服装之外开始穿着西式服装。对于像利伯蒂这样维多利亚时代的人来说，这种变化破坏了他们对东方世界的既有观念。这种观念十分复杂，因为它已经历了长时间的逐步演变，在西方对异质文明的认知和对东方设计的再次解读下成形，这种解读是对东方作为工业化西方国家的对立面的回应。当19世纪末的"日本热"倾向于认为东方世界停滞不前，与西方时装风格的快速变换形成鲜明对比的时候，日本自身正迅速地汲取着西方的影响来改良自己的时装设计。

本土与全球化

20世纪伊始，时装产业因而从这一复杂的历史中逐步发展起来。一方面，某些国家，特别是处于西方对东方笼统概念之下的国度，被视为丰富的感官灵感源泉；而另一方面，西方人通常只把世界其他地区当作一种资源，而非对手。贸易网络几个世纪以来虽然也历经了改变与革新，但常常被西方力量所控制。时装产业拥有全球范围内的贸易链，然而尚未全球化，真正国际化的公司还未形成，世界各地众多国家也没有形成完善的时装体系。这并不是说时装在西方之外的世界里不存在；其他大陆也上演着时装风格的变幻，由本土的审美趣味与社会结构推动。然而，由设计师、制造商创造，以及零售商与媒体推动的周而复始的时装潮流，在20世纪的后半程发展起来。

两次世界大战之间，法国高级时装势力非常强大，驱动着国际时尚潮流。但是，其成功依靠的不仅是个人定制服装的销售，还有其他国家的制造商可以购买与复制的时装设计销售。与此同时，伦敦、纽约等城市也在努力建立自己在时装界的身份，格外关注设计师品牌和时装引导的制造业。这一过程为战后时装产业的加速发展和成长奠定了基础。高级时装依旧迷恋着法式风格，但其他国家也迅速发展出自身的畅销时装特色，尤其在成衣领域。美国就是一个恰当的例子：1930年代至1940年代，美国的时装经常与一种强调统一民族身份的爱国主义神话捆绑在一起推广。到了1950年代早期，尽管在服装设计与元素中继续使用着美国符号，但人们开始更为注重宣传其国际化的时装特质与时尚地位。美国《时尚》杂志充分展示了这一变化，1950年代杂志开始越来越多地刊登世界更多国家的时装设计作品。除了巴黎和伦敦在其评论和广告中长期占据重要版面，来自都柏林、罗马以及马德里的时装系列也在每一季中得到刊载。虽然《时尚》关注的焦点一直是欧洲和西方世界，但这充分展示了对高端时尚地位的渴求是如何蔓延开来的。

随着这些城市逐渐发展成为潮流中心,美国依靠其设计简洁、方便穿着的分体服装以及优雅的礼服等强项站稳了脚跟。这些服装在战后销往更大规模的市场,而最重要的是,牛仔裤与运动服开始在战后统治全球。牛仔裤对各个年龄、性别、种族、阶层而言都很适宜,因而成为推动一种清晰的时尚态度的全球化进程最为显著的因素。虽说牛仔裤未必全然是自发流行起来的时装风格,但它们的影响力不断提升,表达了消费者对于能搭配各种正式和非正式服装,又足以适应个人风格的那种服装的需求。到了21世纪之初,牛仔裤占据了庞大的国际市场,尽管这可能被解读为时装因而也是全球视觉特征的同质化作用,但牛仔裤非常多变,实际上能借助其数不清的排列组合,彰显民族、宗教、亚文化以及个人身份。以巴西为例,马马奥·韦尔德制作出带有闪亮装饰物的贴身牛仔裤,以凸显穿着者的曲线。在日本,牛仔裤成为人们的一种癖好,收藏者努力寻找着稀有的老式李维斯(Levis)牛仔裤,以及像依维斯(Evisu)这样的本土品牌,它推出了印有品牌特有商标的版型宽松的牛仔裤。给牛仔裤带来丰富多样特点的不止是设计师和受热捧的品牌。当牛仔裤的靛蓝色随着多次水洗变得越来越发白,顺着穿着者的身体折出一条条痕迹时,每个人都能创造出自己独一无二的牛仔裤。牛仔裤可以经常根据顾客需求做修改,也可以与二手或新款服装混搭在一起,打造出属于某一特定地域的小规模时装潮流。通过这种方式,人们可以抵抗同质化与全球化,或者至少以自己的创造力赋予它们与本土而非全球推动力相关的全新感觉。

因此,穿着者将自己服装和配饰个性化的过程使得原本全球化对视觉风格影响的简单解读复杂化了。但是，在众多例子中，大品牌在全球的扩张带来了商业街、购物中心以及机场免税店，这些地方几乎全都由相同的品牌构成。像飒拉这样的连锁品牌对本地街头出现的时装潮流能快速做出反应，并将其纳入他们的设计中，这一过程能够使他们在不同国家甚至是不同城市的不同分店中销售不同的产品。不过，在其他情况下，西方品牌对市场的统治可能导致世界各国特定社会阶层的时装风格在视觉上同质化，早期的精英人群中就出现了这种情况。全球的时尚杂志中展示着相同品牌的太阳镜、手袋以及其他配饰，然后被渴望获得所谓的全球高端时尚风格的消费者们纳入囊中。其先驱者很明显是巴黎高级时装自17世纪开始的行业统治，但到了1970年代，各国的"喷气式飞机阶层"出现后，人们渴望的就很容易是意大利或者美国品牌了。许多城市的富人们始终坚持着自己的时装风格，从而催生出依赖社会边界而非地域限制的跨国界时装。

尽管如此，细节上的差异仍然显现出来，比如说，一个民族对美丽与性别的理念上的差异。年龄是影响这些时尚潮流解读方式的另一个重要因素。1990年代，英国品牌博柏利标志性的围巾、军装式风衣以及手袋在韩国青年人群中大受欢迎。尽管我们可以将其看作同质化的一个实例，但品牌标志的格纹却以不一样的方式被穿者演绎着。韩国的情况与日本相同，人们渴望通身都是设计师服装，从鞋子到发饰的所有单品都是大牌。这种显眼的消费在西方看来并不时髦，西方着重于穿着者组合大牌并将其与古着或无名单品混搭的能力，大牌的商标不过是周期性流行一下。韩国青年对博柏利产品的狂热因而颠覆了其内敛英伦上流阶层品味的品牌形象。

玛格丽特·梅纳德指出了加强的时尚潮流国际融合之间的这种复杂的相互作用，认为在一定程度上这是全球化品牌的结果，后者是20世纪末全球变革的产物。她认为，这一现象标志着全球化开始影响经济、政治以及社会生活，因而也会影响时装产业。玛格丽特援引了包括共产主义瓦解、后殖民统治终止、跨国公司与银行发展、全球媒体与网络成长等诸多国际事件，认为它们都是为时尚服装与形象带来大规模传播与流通，促使无数国家时装市场觉醒的原因。国际旅行以及移民方式的不断增加进一步加速了地域边界的消失，以及与之相伴的全球化进程。这一过程也引发了许多道德问题，例如，西方资本主义对廉价制造业的搜刮，再比如，与其同步崛起的快时尚也已见证了自身工业生产的衰退。从古驰这样的奢侈品大牌到盖璞这类大众市场品牌，都将它们的产品制造外包到了中国、越南和菲律宾等国家。这引发了全球化最为罪恶的负面影响——对工人的压榨剥削。如今，追踪供应商并维持工厂标准变得十分困难。工人们一直遭受着用工虐待和薪水压榨，他们还常常来自人口中最为弱势的群体，比如儿童或新来的移民。全球化就这样戴上了一副假面，藏身面具之后的是不公正的工业生产勾当。时装产业巨大的地域覆盖范围使得那些未加入工会的劳动力很容易被雇用到，来为增长中的国际市场提供廉价时装。这也意味着，奢侈品巨头们，比如著名的路易威登集团，如今已然统治了整个行业，除此之外还有一些主打运动服饰和面向年轻群体的品牌，比如迪赛（Diesel）和耐克（Nike）。不过，梅纳德认为，本土差异依旧能够打破全球市场供应商品所带来的潜在大规模同质化，因此完全统一的时装造型或者时尚观念并未在全球范围内造成普遍影响。

塞内加尔国内的时尚潮流就是这种本土形成的流行文化的典型例子，它们一边利用着大规模企业产出的时装大众文化，同时又能够抗拒其影响。塞内加尔年轻人喜欢用来自全球各地的不同风潮丰富自己的衣着风格，并自信地将欧洲与伊斯兰元素以及时装的不同类型整合在一起。尽管牛仔裤和美国黑人风格的流行显而易见，但年轻人仍然会委托本土裁缝们制作出更为正式的款式。胡迪塔·尼娜·穆斯塔法就指出了远在法国殖民之前，塞内加尔就一直很重视个人形象。她详细描写了塞内加尔男女是如何穿着欧非混合时装与本土特有的服饰的。首都达喀尔有一批善于应变的裁缝、制衣师和设计师，包括著名的乌穆·西，她把自己的作品出口到突尼斯、瑞士和法国，这些人对时装进行了成熟的世界性的利用。他们创造出以当下流行的本土风格、传统染色及装饰元素、国际名流，还有法国高级时装为灵感的服装。全球化的贸易网络使得塞内加尔商人能够订购北欧的纺织品设计，收购尼日利亚的织物，然后在欧洲、美洲和中东开展贸易活动。整个国家的时装体系因而整合了本土与全球的潮流，创造出最终到达消费者手里的时装。它很快成为全球化时装产业的一部分，但同时又保留着自身的商业模式与审美品味。达喀尔这座充满活力的时尚都会是21世纪各国时装产业能够共存、共生的典范。确实，就像莱斯利·W.拉宾所说的，整个非洲融合了种类繁多的时装风格与商业类型，它们既在西方资本主义工业体系内运作，又不断探索其边界，"借助那些用手提箱和旅行箱运输货品的手提箱小贩们构成的商业网络，生产者和消费者们创造出跨越国界的流行文化形式"。这样，街头商人（比如早期的小贩）、往返各地的旅行者和观光客，还有长期性甚至永久性的移民人群将不同的时装和配饰传播到全球的各个角落。种种正式和非正式的方式使得原本清晰的民族身份区分变得模糊起来，正如全球化品牌商品传播所带来的结果那样。事实上，这些方式与国际二手服装贸易一起，协力抵抗着这些全球化品牌常常代表的同质化理念。

在欧洲和其他城市中展示的最新时装系列，同样吸收了跨国时装设计理念，融合了十分丰富的文化与民族元素，不再能被任一地理区域明确定义。曼尼什·阿若拉的作品就是这样的例子，因为他把东方与西方的设计和装饰风格结合在了一起。20世纪早期的保罗·普瓦雷等设计师是在西方殖民主义的视角下运用中东与远东的时装元素，阿若拉则摒弃了这种等级观念。不过，西方世界的"东方化"风潮的确深刻影响了视觉与物质文化。关于谁在生产、控制、支配着图像与时装风格使用的问题一直存在。文化借用在时装中广为运用，它为人们的观念构建、风格形成、色彩运用等提供了丰富的跨文化交流。但是，乔斯·突尼辛也提出了她的质疑：

> 异域文化自身的形象常常取决于处于支配地位的西方世界。究竟什么是印度？是印度人民认为的印度，还是我们这些有着殖民统治历史的西方人曾经以为的印度呢？

21世纪之初，这始终是一个令人担忧的议题，考虑到西方悠久且问题重重的殖民统治与支配历史，关于西方设计师运用

"异域"元素是否有所不同的问题也一直悬而未决。或许后现代主义给设计师们对来自众多民族与历史借鉴元素观念的有趣糅合提供了充分的理由,就像我们在约翰·加利亚诺的作品中看到的那样。不过,这并不足以完全抹去时装产业的演变背景,或是这些文化借用的历史含义,以使时装设计与美学趣味抑或时装产业的其他领域(例如贸易交往)实现平等交流。随着越来越多的国家开始在国际上推广自己的时装,这些差异也许会逐渐缩小。在足够多的非西方世界的设计师、奢侈品牌以及成衣制造商拥有与路易酩轩集团及其一众竞争者相当的实力和影响之前,这一过程将会持续下去。

时装周把一个国家或一座城市中的设计师集合起来以展示其每一季的时装系列,它继续提供着一个中心,通过这个中心来宣传某一区域的视觉身份,同时为自己的时装设计师开发并提供平台。时装是一个具有十分重要的经济与文化意义的巨大产业,比如,时装周在众多南美洲城市的传播充分展示了它们如何能够建立起另类的时装中心。1970年代末到1980年代初在巴黎办秀的日本设计师取得了巨大成功,充分证明了非西方设计师也可以给全球市场带来深刻的影响。这一时期,设计师们仍旧需要在知名的时装周办秀以获得足够的知名度和曝光度。山本耀司、川久保玲、高田贤三、三宅一生等诸多日本设计师的作品震惊了西方时装世界,让他们意识到高级时装完全可以发源于自己的界限之外。尤为重要的是,日本时装带来了一种人体与面料以及两者之间相互作用关系的全新视角。

比如三宅一生,他制作的服装颠覆了西方世界关于美与形的固有观念,推出了细密打褶的面料,塑造成向身体外伸出的尖

图20　三宅一生棱角分明的褶皱设计（1990）

端。他重新创造出了符合建筑空间理念的女性气质，不再顺着人体自然形态剪裁面料。其服装形态常常向上延展，并向外延伸以强化身体与服装之间的对比。他的作品被推上国际舞台，在全球各个城市中展示及销售。不过，到了1990年代，三宅曾有过表态，尽管（或者说也许是因为）全球"边界在我们眼前每天被消解又重新定义着……在我看来这是非常必要的。毕竟边界是文化与历史的表达"。他既保持自己的日本身份，同时又能打造出拥有跨越国界的共鸣与魅力的作品，这正是时装产业全球化种种问题的核心。20世纪末以来，贸易网络、商品生产、消费行为以及时装设计，全都越来越紧密地与全球化的时装体系联系在一起。不管对设计者还是穿着者来说，时装的全球化都没有完全压抑本土与个人通过时装表达的东西。但是，21世纪之初的经济衰退或许会加快那些建立在成熟生产模式之上的非西方时装设计的发展，并引发世界时装势力平衡的一次巨变。

结　语

泰莉·艾金斯在她1999年的重要著作《时装的终结》(*The End of Fashion*)里记述了她眼中所见的产业在20世纪末从时装向服装的转变。她认为法国高级时装放慢了脚步,以满足强调价位合理又实穿的经典款式的需求,并且依靠特许经营权来维持生存,尤其是它在全球范围内的香水销售。与此同时,欧洲的大公司们已经发现,比起迪奥的约翰·加利亚诺等风格较为夸张的英国设计师,像赛琳(Celine)麾下的迈克·高仕等美国设计师能为其系列品牌带来更多的销售。艾金斯概述了设计师们对营销而非设计革新的关注。这带来的后果是大众开始对时装审美疲劳,而对盖璞、香蕉共和国(Banana Republic)等高街连锁品牌更感兴趣,因为它们能够可靠地提供各类基本服装款式,偶尔又能引发时尚潮流。艾金斯的论述非常有说服力,它恰好发表于国际经济衰退与远东股市崩溃这十年的尾声。正如她所指出的,由于极简主义设计已经流行,简化的服装本身就是偏离精致时装的潮流的一部分。

那么，时装在1990年代真的终结了吗？这意味着日常的服装最终胜出了吗？艾金斯的确指出了国际市场中一股重要的流行趋势。不过，最有趣的或许是它自身也是一股潮流。正如她自己所说，极简主义在当时是一种时尚，所以市场各个层面中出现的极简风格都是这种时尚的一部分。我们还必须注意到其他潮流也在显现。亚历山大·麦昆等年轻设计师自1990年代初开始崭露头角，创建了既依赖于特许经营权，又在时装设计上不断创新的个人品牌。重要的是1990年代中期，也就是艾金斯认为的开始偏离时装的转折点，正是马修·威廉姆森等设计师在作品中开始对手工技艺与细节制作兴致渐浓的时期。或许艾金斯发现的并非时装的终结，而是时装自始至终灵活多变形式的一个实例。随着文化、社会以及经济环境的演变，设计师的灵感、消费者的需求以及更为关键的欲望，也会不断变化。

诚然，从街头时尚到高级时装，当时掀起了一股强势的以工作服饰为灵感的潮流，它涵盖了各种各样的元素，既有工装裤又有垃圾摇滚风，还有吉尔·桑德等设计师践行的风格冷淡、充满知识分子情怀的极简主义。但是，还须记住的是各种各样的时装风格是同时存在的；当时的高级时装里哥特服饰、暗黑与拜物风格又有所复兴。此外还有威廉姆森水果般色彩鲜艳的时装，把奢华的细节与鲜亮的印花重新带回人们的视野。当美国仍然钟情于盖璞的时候，它在欧洲却开始衰落，盖璞产品宽松的版型和并不鲜明的风格难以与新兴的时髦又令人兴奋的对手们抗衡，比如英国的Topshop和法国的蔻凯（Kookai）。艾金斯因此会写到美式时装、服装品味与生活方式的转折点，而此时正是替代此种现象的新事物吸引大众想象的时刻。她因而确实眼光独到地看到了这一时刻在时装史中的重要性,但是,时装表面上的衰退其实如同黎明前的黑暗时刻,它必将再次勃兴,成为从高街时尚到高级时装的驱动力量。

艾金斯的著述让我们再一次清晰地意识到时装可以不断借鉴吸收外界影响的这种与生俱来的能力,它能够按照有时甚至能通过预期新的生活方式与审美来重塑自己。21世纪之初,服装一如既往地是市场中的重要组分,在艾金斯看来,衣橱经典款式的需求也从未停歇。不过,新兴的高级时装设计师,比如浪凡的阿尔伯·艾尔巴茨、巴黎世家的尼古拉斯·盖斯基埃、圣罗兰的斯特凡诺·皮拉蒂以及巴尔曼(Balmain)的克里斯托夫·狄卡宁,又一次引发了全球对法国时装的热情。即便大多数人只会想要购买他们的标志性手袋,但这些设计师每一季的风格表达很快就出现在高街连锁品牌中。美国的青年设计师们依旧延续着这个国家引领运动服饰的历史,但普罗恩萨·施罗(Proenza Schouler)和罗达特(Rodarte)这些牌子把这些风格转变为奢华的形式,装饰以高级时装的精致细节。伦敦的新兴设计师,比如托德·林恩、路易斯·戈尔丁和克里斯托弗·凯恩则分别重新表现出对精致剪裁、创意十足又色彩丰富的针织衣物以及季节性变化的廓形的兴趣。

世界上的其他城市同样热衷于开发时装这种视觉与物质形式。最典型的应该是印度、中国、南美洲以及太平洋沿岸地区,在那里时装周开始推广本土设计师,同时又积极探寻国内国际不同品牌。在中国,对时装设计教育与促销贸易的兴趣超过了对产能的投资热情,力图在未来打造强大的时装设计形象。印度和俄罗斯新兴的中产阶级和上层阶级意味着这个全新的人群渴望通过服装来表达自己的身份与品味。新的时装杂志涌现出来,既有重要的拥有不同国别版本的《时尚》、《她》与《嘉人》(Marie Claire),也有以本土风格为灵感的各类新刊物。

在街头,时装比以往都更加显而易见,并分门别类地呈现在各大网站页面上,比如http://www.thesartorialist.com,此外也有一些专注介绍从斯德哥尔摩到悉尼等特定城市人群时尚风格的网站。它们充分证明了时装始终具有的通过组合既有潮流与新兴年轻潮流来表达个性的能力。亚文化时装同样充满了活力,这包括对1980年代传遍全球的哥特风的改良,以及与之相关的青少年情绪摇滚(emo)风潮。俱乐部时装风尚越来越艳丽夺目,借鉴了1980年代新浪漫主义与"锐舞"(Rave)文化。一如既往地,时装通过借鉴自己的历史不断向前发展。它交叉借鉴着自己的过去,将重新排列后的风格组合在一起。就这样,克里斯托弗·凯恩以阿瑟丁·阿拉亚1980年代的紧身裙与范思哲1990年代初充满活力的设计为灵感,创造出令人耳目一新的时装。"新锐舞"重新改造了上一代的荧光色与超大码标语T恤衫。在这些例子中,新世纪见证了人们对大小与色彩的喜好,而1990年代的大部分时装都缺乏这些。

21世纪之初也见证了越来越多具有道德诉求的品牌和网站的建立,它们关注时装对地球产生的影响,同时关心工人的权益。在各类报道曝光了从墨西哥到印度等国家为西方知名品牌生产服装的工厂剥削现象后,这些品牌和网站的兴起代表了对此种现象的有力回应。时装开始需要处理自己的生产方式,这是一个重要的转变。尽管从19世纪中叶开始人们就在呼吁,但对此的回应却是时断时续。道德时装的此次繁荣能否渗入整个产业，为纺织品制造与服装生产方式带来永久、深远的变革，仍需拭目以待。我们希望这是一股长期的潮流趋势，而非昙花一现的风尚。

与此同时，时装也成长为学术研究的对象，有越来越多的专著和期刊来研究其本质、地位以及意义。全球各大博物馆推出的时装展览大获好评，引发人们对时装的极大热情。在市场的另一端，名流文化的兴起使得时装的传播速度甚至超越了好莱坞的全盛时期。社会、文化以及政治生活方式与态度的这些不同方面，逐渐与时装的诞生、传播以及越来越全球化的特性联系在一起。时装因而并未终结，但它确实发生了变化，并且极有可能处在另一次深刻变革的边缘。随着非西方时装体系暗自发展壮大，经济衰退又席卷而来，时装主力很可能会转向东方。尽管自文艺复兴时期演化而来的西方时装产业不太可能被其吸纳，但面对来自全球的挑战，它必须学会快速适应并做出精准有效的回应。

索 引

（条目后的数字为原书页码，见本书边码）

A

A Magazine《A 杂志》62
A. T. Stewart, New York 纽约的斯图尔特 73
Academy Awards 美国电影艺术与科学学院奖 61
accessories 配饰 10, 11, 12, 19, 31, 35, 51, 67, 70, 74, 76, 77, 81, 82, 94, 117, 120
advertising 广告 3, 14, 18, 23, 32, 50, 55, 60—61, 62, 66, 68, 69, 72, 75, 77, 78, 79, 83, 87, 88, 89, 97, 116
Aesthetic dress 唯美服装 40, 75
Aganovitch and Yung 阿加诺维奇与杨 108
Agins, Teri 泰莉·艾金斯 57, 124—126
Agnes B. 阿尼亚斯·B. 44
Alaia, Azzedine 阿瑟丁·阿拉亚 127
Alphabet City, New York 纽约字母城 81
America 美国 7, 10, 13, 15, 17, 18, 19, 20, 21, 23, 41, 43, 52, 53, 54, 57, 62, 75, 76, 77, 78, 82, 85, 94, 95, 99, 100, 101, 109, 110, 113, 116, 118, 120, 124—126
American Apparel AA 美国服饰 89—90, 101
American Civil War 美国内战 53
American fashion 美国时装，见 America
Amies, Hardy 赫迪·雅曼 99

Another Magazine《新杂志》62
Antwerp 安特卫普 1—2, 22, 62
Apraxine, Pierre 皮埃尔·阿普拉克西纳 40
Argentina 阿根廷 77
Arletty 阿莱缇 95
Armani, Giorgio 乔治·阿玛尼 24
Armstrong, Lisa 莉萨·阿姆斯特朗 107
Arnold, Janet 珍妮特·阿诺德 5
Arora, Manish 曼尼什·阿若拉 105—108, 120
Asia 亚洲 56, 71, 97, 110—111, 114—115
Asome, Caroline 卡罗琳·阿索姆 108
Austen, Jane 简·奥斯汀 59
Australia 澳大利亚 77

B

Bailey, Christopher 克里斯托弗·贝利 84
Baillén, Claude 克洛德·巴扬 10
Balenciaga 巴黎世家（巴伦西亚加）33—34, 66, 126
Ballet Russes "俄罗斯芭蕾" 41
Balmain 巴尔曼 126
Banana Republic 香蕉共和国 124
Barcelona 巴塞罗那 67, 114
Barney's, New York 纽约巴尼斯精品时装专卖店 83
Barthes, Roland 罗兰·巴特 5
Bauhaus 包豪斯派 44
Bayer, Herbert 赫伯特·拜耶 44
Bazaar, London 玛丽·奎恩特伦敦

"集市"店 80
Beaton, Cecil 塞西尔·比顿 17
Belfanti, Carlo Marco 卡洛·马可·贝尔凡蒂 112
Benetton 贝纳通 69
Benson, Susan Porter 苏珊·波特·本森 74
Berlin 柏林 67
Bershka 巴适卡 81
Bertin, Rose 罗斯·贝尔坦 12, 13
Bextor, Sophie Ellis 苏菲·埃利斯·贝克斯特 85, 86
Birth of a Dress, The《一条裙子的诞生》48—50, 49
Bon Marché 乐蓬马歇商店 73, 75
Bond Street, London 伦敦邦德街 74
Boston 波士顿 74
Boucicaut, Astride 阿里斯蒂德·布西科 73
boutiques 精品店 13, 20, 24, 31, 44, 47, 67, 69, 70, 78, 79, 80, 81, 83
brands 品牌 7, 50, 56, 58, 69, 75, 77, 82, 83, 84, 89, 90, 96, 100, 101—103, 107, 117—121, 124, 127
Breakfast at Tiffany's《蒂凡尼的早餐》61
Breward, Christopher 克里斯托弗·布鲁沃德 5
Brigg Market, Leeds 利兹布里格市场 71
Brixton, London 伦敦布里克斯顿 75
Broadway, New York 纽约百老汇 73
Brodovitch, Alexey 阿列克谢·布罗多维奇 62
Bruges 布鲁日 71

Burberry 博柏利 7, 84, 118
Burgundian Court 勃艮第宫廷 51
Burton's 博尔顿男装 57
buyers 买手 8, 49, 50, 55, 59, 78, 108, 122

C

Campbell, Naomi 娜奥米·坎贝尔 87
Cardin, Pierre 皮尔·卡丹 20
Carter, Ernestine 欧内斯廷·卡特 10
Cashin, Bonnie 邦妮·卡欣 21, 23
Castiglione, Virginia Versasis, Countess de 卡斯蒂廖内伯爵夫人，维尔吉尼娅·维拉西斯 38—40
Casuals 休闲派 26
Celine 赛琳 124
Cepea Fabrics 西帕纺织公司 48
Cerruti, Nino 尼诺·切瑞蒂 24
Cézanne, Paul 保罗·塞尚 38
chain store 连锁店 77, 81, 82, 101, 102, 117, 124, 126
Chambre Syndicale de la Haute Couture 巴黎高级时装公会 20, 52
Chanel 香奈儿 8—11, 18, 19
Charney, Dov 多夫·查尼 89
Cheap Date《邂逅》60
Chicago 芝加哥 76
China 中国 22, 84, 107, 108, 109, 112,113, 119, 126
Chloé 蔻依 11
Cho, Margaret 玛格丽特·曹 97
Christina of Denmark 丹麦的克里斯蒂娜 35
Clark, Judith 朱迪思·克拉克 1—2

Clark, Larry 拉里·克拉克 89
Clark, Michael 迈克尔·克拉克 22
Cocteau, Jean 让·科克托 42
colleges 院校 5, 22, 79, 107
Collins, Kenneth 肯尼斯·柯林斯 78
Comme des Garçons CDG 22, 44, 67—70, 68, 123
Commedia del Arte 即兴喜剧 1
Constructivism 构成主义 43
Costume Institute, Metropolitan Museum, New York 纽约大都会艺术博物馆服饰馆 5
Couturier 高级时装设计师 10, 12—21, 24, 38—43, 48, 57, 60, 61, 66, 70, 78, 82, 126
Covent Garden 考文特花园 72
craftsmanship 工艺 4, 19, 105
Cranach, Lucas 卢卡斯·克拉纳赫 37
Crawford, Joan 琼·克劳馥 61
Cruikshank, George 乔治·克鲁克香克 92
Czechoslovakia 捷克斯洛伐克 41, 99

D

Dakar 达喀尔 119—120
Dali, Salvador 萨尔瓦多·达利 42
Decarnin, Christoph 克里斯托夫·狄卡宁 126
Degas, Edgar 埃德加·德加 38
Demarge, Xavier 格扎维埃·德马尔涅 40
Demeulemeester, Ann 安·迪穆拉米斯特 22
department store 百货公司 15, 20, 21, 44, 55, 70, 73—79, 114
Dickens, Charles 查尔斯·狄更斯 15
Diesel 迪赛 119
diffusion lines 品牌副线 57
Dior, Christian 克里斯汀·迪奥 1, 19, 66, 124
Dior Homme 迪奥·桀傲男装 26
Disney, Walt 华特·迪士尼 105
DKNY 唐可娜儿 57
Dover Street Market "丹佛街集市" 69—70
dressmakers 制衣师 12, 14, 17, 18, 59, 60, 69, 70, 76, 120
Drottningholm Palace 皇后岛夏宫 113
Dublin 都柏林 116
Dudes "时髦男" 94
Dufy, Raoul 劳尔·杜飞 42
Dunn, Jourdan 乔丹·邓恩 97

E

East India Company 东印度公司 71, 111
Elbaz, Alber 阿尔伯·艾尔巴茨 19, 126
Elle《她》62, 126
Eloffe, Madame 埃洛弗夫人 13
England 英国 17, 49, 70, 94, 111, 114
Englishwoman's Domestic Magazine, The《英国女性居家杂志》60
Entwistle, Joanne 乔安妮·恩特威斯尔 5
Epicoene, or The Silent Woman《艾碧

辛，或沉默的女人》91
Esprit 埃斯普利特 7
Europe 欧洲 12, 13, 19, 33—34, 70, 97, 99, 108—113, 115, 118, 119, 123, 124
Evans, Caroline 卡罗琳·埃文斯 2, 5
Evisu 依维斯 117

F

fabric 面料 11, 12, 24, 26, 33, 35, 36, 37, 42, 48, 49, 51, 52, 53, 55, 64, 70, 71, 72, 73, 75, 88, 90, 98, 99, 101, 109, 111, 113, 114, 120, 123
Face, The《面孔》60
Far East 远东地区 20, 83, 96, 100, 103, 109, 110, 120, 124
Fashion Design Council of India 印度时装设计理事会 107
fashion shows 时装秀 2, 3, 8, 22, 32, 33, 55, 58, 59, 66, 76, 78, 108
Fashion TV 法国时尚电视台 59
Fast Fashion 快时尚 56, 100, 119
Fath, Jacques 雅克·法特 20
Fendi 芬迪 11
Fifth Element, The《第五元素》61
Filene, Edward 爱德华·法林 74
flagship stores 旗舰店 82, 83
Florence 佛罗伦萨 51, 109
Foale and Tuffin 福阿莱与图芬 31
Fondazione Prada 普拉达基金会 45
Fops "花花公子" 93
Ford, Tom 汤姆·福特 84
forecasting 预测 20, 50
France 法国 6, 13—14, 17, 20—21, 51, 52, 57, 71, 76, 78, 94, 95, 96, 97, 99, 109, 114,
119—120, 124, 125, 126
Frankel, Susannah 苏珊娜·弗兰克尔 23
French fashion 法国时装, 见 France
French Revolution 法国大革命 13, 94
Frissell, Toni 托尼·弗里塞尔 64—65
Future Systems "未来系统" 44
Futurism 未来主义 42

G

Galliano, John 约翰·加利亚诺 1, 19, 65—66, 121, 124
Gap 盖璞 56, 69, 101, 119, 124
Gas Council 英国煤气委员会 48
Gaultier, Jean-Paul 让-保罗·高缇耶 57, 61
Geneva 日内瓦 71
Germany 德国 44, 99
Ghesquiere, Nicolas 尼古拉斯·盖斯基埃 126
Gimbel, Sophie 苏菲·金贝尔 79
Givenchy 纪梵希 57, 61
Gn, Andrew 安德鲁·鄞 108
Godley, Andrew 安德鲁·戈德利 50
Goldin, Louise 路易斯·戈尔丁 126
Goldin, Nan 南·戈尔丁 89
Goubard, Madame Marie 玛丽·古博夫人 60
Goya 戈雅 37
Grand Mogul, Paris 巴黎"大人物"精品店 13
grande Pandora "大潘多拉" 13
Great Depression 大萧条时期 21, 22,

56, 79
Grès, Mme 格雷夫人 3
Guatemala 危地马拉 114
Gucci 古驰 83, 84, 119
guerrilla store 游击店 67—68
Guggenheim Museum, New York 纽约古根海姆美术馆 44
Gupta, Subdodh 苏伯德·古普塔 105
Gursky, Andreas 安德烈亚斯·古尔斯基 45
Gustav III 古斯塔夫三世 113

H

H&M 海恩斯莫里斯 56, 82, 102
Halpert, Joseph 约瑟夫·哈尔珀特 20
Hamilton, Alan 艾伦·汉密尔顿 108
Harajuku street fashions 原宿街头时尚 26, 27, 96—97; 参见 Japan
Haring, Keith 凯斯·哈林 43
Harper's Bazaar 《哈珀芭莎》62, 79
Harrods, London 伦敦哈罗德百货 76
Hartnell, Norman 诺曼·哈特内尔 18
haute couture 高级时装 6, 10, 18—21, 40—41, 48, 52, 116
Hawaii 夏威夷 83
Hebdige, Dick 迪克·赫伯迪格 5
Hebrew 希伯来 109
Hello 《问候》61
Henry VIII 亨利八世 35
Hepburn, Audrey 奥黛丽·赫本 61
Hepworth & Son 赫普沃斯公司 77
high street 高街时尚 4, 26, 27, 56, 66, 69, 81—83, 100, 117, 124, 125, 126
Hilliard, Nicholas 尼古拉斯·希利亚德 35
Hippies 嬉皮士 100
Holbein, Hans 汉斯·霍尔拜因 35—36
Holland 荷兰 111
Hollander, Anne 安妮·霍兰德 5, 37
Hollywood 好莱坞 24, 31, 61, 77, 95, 127
Humanism 人文主义 6, 34

I

I. Miller I. 米勒制鞋 30
identity 身份 5, 15, 17, 18, 26, 31, 32, 35, 40, 45, 47, 52, 60, 69, 70, 72, 75, 77, 92, 96, 97, 112, 116, 117, 120, 121, 123
Ilinic, Roksanda 洛克山达·埃琳西克 108
illustration 插画 14, 17, 63—64, 72
India 印度 1, 22, 84, 100, 105—109, 110—111, 112, 113—114, 121, 126, 127
Inditex 印第纺集团 81—82
individuality 个性 3, 6, 26, 34, 38, 51, 53, 69, 92, 117, 123, 126
Industrial Revolution 工业革命 6
Islam 伊斯兰 97, 109

J

Jacobs, Marc 马克·雅各布 57, 84
jacquard loom 提花织机 53
Jaeger, Dr Gustav 古斯塔夫·耶格尔博士 99
Japan 日本 22, 26, 27, 44, 57, 83, 96—97,

135

102, 105, 108, 112, 115, 117, 118, 122—123

Japanese fashion 日本时装，见 Japan

jewellery 珠宝 8, 10, 35, 76

Joffe, Adrian 阿德里安·约菲 69

Johnson, Betsey 贝齐·约翰逊 31

Jones, Ann Rosalind 安·罗莎琳德·琼斯 112

Jones, Inigo 伊尼戈·琼斯 72

Jonson, Ben 本·琼森 91

Josephine, Empress 约瑟芬皇后 14

Junior Gaultier 高缇耶童装 57

K

Kane, Christopher 克里斯托弗·凯恩 58, 126

Karan, Donna 唐娜·凯伦 10

Kawakubo, Rei 川久保玲，见 Comme des Garçons

Kendel Milne, Manchester 曼彻斯特的肯德尔·米尔恩 73

Kenzo 凯卓 57, 123

Kershen, Anne 安妮·科尔申 50

Kidwell, Claudia 克劳迪娅·基德韦尔 53

Killerby, Catherine Kovesi 凯瑟琳·克韦希·基勒比 98

King's Road, London 伦敦国王大街 80

Klein, Calvin 卡尔文·克莱恩 23, 89

Kookai 蔻凯 125

Koolhaas, Rem 雷姆·库哈斯 45

Kors, Michael 迈克·高仕 124

Kyoto Museum 京都博物馆 5

L

Lady's Magazine, The《淑女杂志》60

Lagerfeld, Karl 卡尔·拉格斐 8—11, 19, 82

Lambert, Eleanor 埃莉诺·兰伯特 60—61

Lanvin 浪凡 19, 126

Laurent, Yves Saint 伊夫·圣罗兰 20, 24, 84

Leipzig 莱比锡 71

Lelong, Lucien 吕西安·勒隆 20

Lemire, Beverly 贝弗利·勒米尔 5, 51

Lepape, Georges 乔治·勒帕普 16

Leroy, Louis Hyppolite 路易·伊波利特·勒罗伊 14

Let it Rock, London 伦敦"尽情摇滚" 80

Levi Strauss 李维斯 53, 117

Liberty 伦敦自由百货 73—74, 113

Liberty, Arthur Lasenby 亚瑟·莱森比·利伯蒂 114—115

lifestyle 生活方式 10, 23, 62, 69, 74, 125, 127

Limnander, Armand 阿曼德·利姆南德尔 102

London 伦敦 12, 13, 18, 24, 44, 48, 52, 54, 58, 61, 69, 72, 74, 75, 76, 80, 97, 107, 108, 114, 116, 126

Lord & Taylor 罗德泰勒百货公司 21, 44

Los Angeles 洛杉矶 89

Louis Vuitton 路易威登 45, 57, 84

Louis Vuitton Moët Hennessey 路威

酩轩 57, 121
Louis XIV 路易十四 52
Lucile 露西尔 6—7, 17, 55
luxury brands 奢侈品牌 19, 57, 83, 84, 103, 107, 117, 119, 121
Lynn, Todd 托德·林恩 126
Lyons 里昂 52

M

MAC 魅可 108
Macaronis "纨绔子弟" 93—94
Macy's, New York 纽约梅西百货 78
Madonna 麦当娜 82
Madrid 马德里 116
magazines 杂志 4, 6, 13, 31, 40, 50, 59—62, 71, 72, 79, 84, 90, 98, 117, 126; 参见具体名称
Malign Muses "恶毒的缪斯" 1, 1—4
Mamao Verde 马马奥·韦尔德 117
Man About Town 《都会男士》62
mannequins 模特 6, 33; 参见 models
marchandes des modes "款式商人" 12—13
Margiela, Martin 马丁·马吉拉 23, 62, 102
Marie Antoinette 玛丽·安托瓦内特 12
markets 市场 60, 70—71, 72, 83, 102, 103
Marks and Spencer 玛莎 102
Marshall and Snelgrove, Scarborough 马歇尔和斯奈尔斯格罗夫百货斯卡伯勒分店 75
Marshall Ward, Chicago 芝加哥马歇

尔沃德百货 77
Mashers "花花公子" 94
mass-market 大众市场 82, 84, 101, 119
mass-produced 大批量生产 20, 43, 48—50, 52—56, 77, 80, 83, 100
Massimo Dutti 马西莫·杜蒂 81
Matisse, Henri 亨利·马蒂斯 41, 42
Maynard, Margaret 玛格丽特·梅纳德 118
McCardell, Claire 克莱尔·麦卡德尔 43, 61
McCartney, Stella 斯特拉·麦卡特尼 82, 102
McDougall, Dan 丹·麦克杜格尔 101
McLaren, Malcolm 马尔科姆·麦克拉伦 80
McQueen, Alexander 亚历山大·麦昆 1, 22, 125
Mediterranean 地中海风格 76, 109
Mendes, Eva 伊娃·门德斯 87
menswear 男装 12, 24—26, 53, 57, 77, 78, 92
Mexico 墨西哥 52, 113, 127
Middle East 中东 19, 71, 109, 110, 120
Miller, Daniel 丹尼尔·米勒 5
Minogue, Kylie 凯莉·米洛 82
Miyake, Issey 三宅一生 122, 122—123
Mizrahi, Isaac 艾萨克·麦兹拉西 57
Mode Museum, Antwerp 安特卫普时尚博物馆 1, 2
models 模特 8, 22, 23, 24, 33, 42, 59, 64, 65, 76, 81, 82, 87, 88, 89, 97—98, 105
Mods "摩登族" 26, 96
Monet, Claude 克劳德·莫奈 38
Montgolfier brothers 孟高尔费兄弟 13

索引

137

Moore, Henry 亨利·摩尔 44
Moore, Julianne 朱莉安·摩尔 24
Morocco 摩洛哥 41
Moses, Elias 伊莱亚斯·摩西 54
Moss, Kate 凯特·莫斯 23, 82—83
Mouillard, Madame 穆亚尔德夫人 13
Mr Fish 费什先生 24
Muji 无印良品 102
Muslim 穆斯林 97
Mustafa, Hudita Nina 胡迪塔·尼娜·穆斯塔法 119
Myra's Journal of Dress and Fashion《米拉衣装志》60

N

Nagrath, Sumati 须摩提·纳格拉斯 108
Napoleon 拿破仑 14, 52, 76
National Institute of Fashion Technology, New Delhi 新德里国家服装设计学院 107
Net-a-Porter.com 颇特女士网 84
New Look 纽洛克 82, 102
New York 纽约 18, 21, 44, 45, 47, 73, 76, 81, 83, 100, 116
New York Times, The《纽约时报》89, 102
Newcastle 纽卡斯尔 76
Next 奈克斯特 77
Niessen, Sandra 桑德拉·尼森 114
Nigeria 尼日利亚 108, 120
Nike 耐克 119
Northampton 北安普顿 49

Noten, Dries Van 德赖斯·范诺顿 1, 22, 24

O

Observer, The《观察家报》100—101
Obsession "激情" 香水 23
Oliphant, Margaret 玛格丽特·奥利芬特 38
Olowu, Duro 杜罗·奥罗伍 108

P

Pacific Rim 太平洋沿岸地区 126
Palais Royal 巴黎皇家宫殿 72
Paquin, Jeanne 让娜·帕坎 17
Paraphernalia "随身用品" 店 31
Paris 巴黎 7, 13—21, 24, 44, 52, 53, 55, 60, 71, 72, 73, 76, 78, 79, 95, 102, 107, 108, 109, 116, 118, 122
Parisian fashion 巴黎时装,见 Paris
People Tree 英国品牌 "人树" 101
perfume 香水 10, 18, 19, 23, 58, 67, 124
Perkin, William 威廉·珀金 114
Perrot, Philippe 菲利普·佩罗 53
Persia 波斯 112
PETA (People for the Ethical Treatment of Animals) 善待动物组织 85—88
Philadelphia 费城 53, 76
Philippines 菲律宾 119
Picasso, Pablo 巴勃罗·毕加索 43
Pierson, Pierre-Louis 皮埃尔-路易·皮尔森 38
Pilati, Stefano 斯特凡诺·皮拉蒂 24, 126

Pilatte, Charles 查尔斯·皮拉特 14
Poiret, Paul 保罗·普瓦雷 16, 17, 41—43, 120
Poole, Henry 亨利·普尔 12
popular culture 流行文化 11, 20, 31, 41, 119
'pop-up' shops "快闪"店 67
Portman, Natalie 娜塔莉·波特曼 24
Portugal 葡萄牙 110
Posen, Zac 扎克·珀森 24
Powerhouse Museum, Sydney 悉尼动力博物馆 5
Prada 普拉达 24, 45, 45—46
Prague 布拉格 76
Pre-Raphaelites 前拉斐尔派 40
Primark 普里马克 100—101
Prince, Richard 理查德·普林斯 45
Printemps 巴黎春天百货 20
Proenza Schouler 普罗恩萨·施罗 126
Pucci 璞琪 21, 57
Pugh, Gareth 加勒斯·普 58
Punch《笨拙》94
Punks 朋克族 26, 80

Q

Quant, Mary 玛丽·奎恩特 21, 80—81
Queen, The《皇后》62

R

Rabine, Leslie W. 莱斯利·W. 拉宾 120
Rappaport, Erika 埃丽卡·拉帕波特 74
rationing 定量配给 79, 99
Rawsthorn, Alison 艾莉森·罗斯索恩 20
ready-to-wear 成衣 4, 11, 19, 20, 21, 43, 48, 49, 51—57, 65, 66, 72, 75, 77, 78, 79, 82, 83, 107, 109, 116, 121
readymade 成品服装，见 ready-to-wear
Redfern, John 约翰·雷德芬 17
Reebok 锐步 107
Reformation, the 宗教改革 34
Regent's Street, London 伦敦摄政街 114
Renaissance 文艺复兴 6, 34, 51, 59, 70, 72, 98, 110, 128
Respect 英国"尊重"组织 88
Reynolds, Joshua 约书亚·雷诺兹 36—37
Ribeiro, Aileen 艾琳·里贝罗 5, 38, 113
Rive Gauche "塞纳左岸"精品店 20
Roche, Daniel 丹尼尔·罗什 71
Rodarte 罗达特 126
Rome 罗马 19, 116
Rowlandson, Thomas 托马斯·罗兰森 92
Rubens, Peter Paul 彼得·保罗·鲁本斯 37
Ruby London 鲁比伦敦 101
Russia 俄罗斯 12, 19, 43, 54, 62, 84, 126

S

Saks Fifth Avenue 萨克斯第五大道精

索引

139

品百货 21, 79
Salone Moderne "现代沙龙" 79
Sander, Jil 吉尔·桑德 102, 125
Sargent, John Singer 约翰·辛格尔·萨金特 40
Sato, Tomoko 佐藤知子 115
Savile Row 伦敦萨尔维街 12, 24
Scandinavia 斯堪的纳维亚半岛 100
Schapiro, Raphael 拉斐尔·夏皮罗 50
Schiaparelli, Elsa 艾尔莎·夏帕瑞丽 1, 18, 42, 43, 78
Schwab, Marios 马里奥斯·施瓦布 58
Second World War 第二次世界大战 21, 79, 98, 109
second-hand 二手衣物 51, 72, 102, 117, 120
Seditionaries, London 伦敦"煽动分子"精品店 80
Selfridge's 塞尔福里奇百货 75
Senegal 塞内加尔 119—120
Settle, Alison 艾莉森·赛特尔 19—20
Sex, London 伦敦"性"精品店 80
Shand, Dennis 丹尼斯·尚德 48
Shanghai, Peace Hotel 上海和平饭店 46
Shaver, Dorothy 多萝西·谢弗 21
Sherard, Michael 迈克尔·谢拉德 48—49
Sherman, Cindy 辛迪·雪曼 44
Shirley, Robert 罗伯特·舍利 112
shops 商店 12, 29, 45, 47, 56, 89, 91, 102—103, 117; 参见 boutiques
Simons, Raf 拉夫·西蒙 26, 102
Singapore 新加坡 67, 108
Singer 胜家公司 53

Skinheads "光头党" 96
Slimane, Hedi 艾迪·斯里曼 25—26
Slow Fashion 慢时尚 102
Smith, Woodruff D. 伍德拉夫·D.史密斯 71
Smollet, Tobias 托拜厄斯·斯摩莱特 72
Snow, Carmel 卡梅尔·斯诺 62
Society of French Interior Designers 法国室内设计师协会 44
Soho, New York 纽约苏荷区 45
South America 南美洲 56, 57, 100, 113, 121, 126
Soviet Union 苏联 43, 67, 99
Spain 西班牙 56, 110, 113
Spectator, The《观察家报》92
Spectres: When Fashion Turns Back "魅影：时装回眸" 2
spinning jenny 珍妮纺纱机 53
Sri Lanka 斯里兰卡 100
Stallybrass, Peter 彼得·斯塔利布拉斯 112
Stefani, Gwen 格温·史蒂芬妮 96—97
Stepanova, Vavara 瓦瓦拉·史蒂潘诺娃 43
Stiebel, Victor 维克多·斯蒂贝尔 18
Stockholm 斯德哥尔摩 76, 126
storewindows 商店橱窗 21, 48, 66, 71, 73, 77, 78
street fashion 街头时尚 26, 27, 44, 56, 62, 69, 80, 96, 125, 126
Styles, John 约翰·斯戴尔斯 111
subculture 亚文化 26—28, 68—69, 80—81, 95—96, 117

sumptuary laws 禁奢令 98, 113
Swanson, Carl 卡尔·斯旺森 47
Swarovski 施华洛世奇 105
Swatch 斯沃琪 108
sweatshops 血汗工厂 55, 89, 100—101
Sweden 瑞典 56, 101, 113
Swells "时髦人士" 94
Switzerland 瑞士 113
Sy, Oumou 乌穆·西 120
Sydney 悉尼 5, 126
Syria 叙利亚 110

T

Tailor and Cutter: A Trade Journal and Index of Fashion, The《裁缝与裁工：时装行业刊物与索引》62
tailors 裁缝 12, 51—54, 62, 64, 76, 77, 119, 126
Tanaami, Keiichi 田明网敬一 105
Target 美国连锁商店塔吉特 82
Teddy Boys "不良少年" 4, 80
Teunissen, José 乔斯·突尼辛 121
textiles 纺织品 35, 42, 43, 48, 50, 51, 52, 53, 71, 102, 107, 109, 110—111, 113—114, 120, 123, 127; 参见 fabric
Thayaht, Ernesto 欧内斯托·塔亚特 42
They Shoot Horses Don't They?《孤注一掷》22
Thirty Years War "三十年战争" 52
Thomas, Dana 达娜·托马斯 84
Titian 提香 35
Tokyo 东京 22, 44, 96

Toledo, Ruben 鲁本·托莱多 1
Townley 汤利制衣 21
Train, Susan 苏珊·特雷恩 20
Triangle Shirtwaist Factory 纽约三角大楼制衣厂 100
Troy, Nancy 南希·特洛伊 41, 43
Tunisia 突尼斯 120
Turkey 土耳其 110

U

United States Army Clothing Establishment 美国军队服装公司 53
Utility Scheme 英国公共事业计划 99

V

Valentina 华伦蒂娜 18
Valentino 华伦天奴 19
Van Dyck, Anthony 安东尼·凡·戴克 36
Velvet Underground "地下丝绒" 乐队 31
Venice 威尼斯 109
Versace 范思哲 10
Victoria and Albert Museum, London 伦敦维多利亚和阿尔伯特博物馆 2
Victorine 维多琳 14
Vietnam 越南 119
Viktor and Rolf 维果罗夫（维克多与罗尔夫）3, 32—34, 82
vintage 古着 26, 60, 82, 83, 102, 117, 118
Vionnet, Madeleine 玛德琳·维奥内 18, 42, 78
Vittu, Françoise Tetart 弗朗索瓦丝·

索引

泰尔塔・维蒂 14

Vogue《时尚》62, 79, 97, 116, 126

W

Waist Down: Miuccia Prada, Art and Creativity "腰肢以下：缪西娅・普拉达、艺术与创造力" 45—46

Wanamaker's, Philadelphia 费城沃纳梅克百货 76

Warhol, Andy 安迪・沃霍尔 29—31, 34

Warsaw 华沙 44, 67—68

Watanabe, Toshio 渡边俊夫 115

Westwood, Vivienne 薇薇恩・韦斯特伍德 24, 43, 80

WGSN 世界全球潮流互联网 62

Wiener Werkstätte 维也纳工坊 44

Williams, Beryl 贝丽尔・威廉姆斯 23

Williamson, Matthew 马修・威廉姆森 125

Wilson, Elizabeth 伊丽莎白・威尔逊 5

Winterhalter 温特哈尔特 39

Wolf, Jaime 杰米・沃尔夫 89

Wollen, Peter 彼得・沃伦 44

World's End "世界尽头" 精品店 80

Worth, Charles Frederick 查尔斯・弗雷德里克・沃思 15, 38, 55

Wortley-Montague, Lady Mary 玛丽・沃特利-蒙塔古夫人 112

X

XULY Bët 克叙里・比约特 102

Y

Yamamoto, Yohji 山本耀司 22

Z

Zara 飒拉 56, 81, 81—82, 117

Zara Home 飒拉家居 82

Zazous "先锋派" 95—96

时装

Rebecca Arnold

FASHION
A Very Short Introduction

For Adrian

Contents

Acknowledgements i

List of illustrations iii

Introduction 1

1 Designers 8

2 Art 29

3 Industry 48

4 Shopping 67

5 Ethics 85

6 Globalization 105

Conclusion 124

References 129

Further reading 135

Acknowledgements

I would like to thank Andrea Keegan, my editor at Oxford University Press, for her support and encouragement of this project. Thanks to all my colleagues and the students of the History of Design Department at the Royal College of Art in London. I am indebted to Caroline Evans for her excellent advice, and to Charlotte Ashby and Beatrice Behlen for their thoughtful comments on drafts. Thank you to Alison Toplis, Judith Clark, and Elizabeth Currie for their helpful suggestions. And finally, thanks to my family, and to Adrian Garvey for everything.

List of illustrations

1. *Malign Muses* tableau, Mode Museum, Antwerp **2**
   Courtesy of ModeMuseum, Antwerp

2. Chanel couture show, spring 2008 **9**
   © Patrick Kovarik/AFP/Getty Images

3. Georges Lepape, Paul Poiret fashion illustration, 1911 **16**
   V&A Images, The Victoria and Albert Museum

4. Hedi Slimane for Dior Homme spring 2005 **25**
   Courtesy of Christian Dior

5. Japanese Harajuku street fashion **27**
   © Jerry Driendl/Taxi Japan/Getty Images

6. Andy Warhol, *Diamond Dust Shoes*, 1980–1 **30**
   © The Andy Warhol Foundation for the Visual Arts, Inc./ARS, New York/DACS, London 2009. Photo: © Art Resource, New York

7. Pierre–Louis Pierson, Countess Castiglione, c. 1863–66 **39**
   © The Metropolitan Museum of Art, New York/Scala, Florence

8. Prada's *Waist Down: Miuccia Prada, Art and Creativity* exhibition **46**
   © Mario Anzuoni/Reuters/Corbis

9. Still from *Birth of a Dress*, 1954 **49**
   BFI Collections

10. 'Ascension de Madame Garnerin, le 28 mars 1802' **63**
    Courtesy of the Library of Congress

11. Toni Frissell, February 1947 **65**
    Courtesy of the Library of Congress

12. Comme des Garçons guerrilla store in Warsaw, 2008 **68**
    Courtesy of Comme des Garçons SAS

13 Engraving of London clothes market, early 19th century **72**
© Southampton City Art Gallery/The Bridgeman Art Library

14 Zara shop window **81**
© Inditex

15 'Here's the Rest of Your Fur Coat', PETA (People for the Ethical Treatment of Animals) poster, 2007 **86**
Courtesy of PETA

16 Engraving of a Macaroni, 1782 **93**
© The Print Collector/HIP/TopFoto.co.uk

17 Counterfeit luxury brands market in the Far East **103**
© Emma Sklar/Sinopix

18 Manish Arora autumn/winter 2008–9 **106**
Courtesy of Manish Arora

19 Textile, Venice, 1450 **110**
© V&A Museum/Erich Lessing/akg-images

20 Issey Miyake dress, 1990 **122**
V&A Images, The Victoria and Albert Museum/© Issey Miyahe,

Introduction

Malign Muses, Judith Clark's groundbreaking 2005 exhibition at the Mode Museum in Antwerp, brought together recent and historical dress in a spectacular series of tableaux. The setting was designed to look like a 19th-century fairground, with simple plain wooden structures that evoked carousels, and oversized black and white fashion drawings by Ruben Toledo, which added to the feeling of magic and showmanship. The exhibition emphasized fashion's excitement and spectacle. Intricate designs by John Galliano and Alexander McQueen mixed with interwar couture, including Elsa Schiaparelli's 'skeleton dress', a black sheath embellished with a padded bone structure. A dramatic 1950s Christian Dior evening dress in crisp silk, with a structured bodice and sweeping skirt, caught with a bow at the back, was shown, as was a delicate white muslin summer dress made in India in the late 19th century, and decorated with traditional chain stitch embroidery. Belgian designer Dries Van Noten's jewel-coloured prints and burnished sequins of the late 1990s stood next to a vibrantly hued Christian Lacroix ensemble of the 1980s. This extravagant combination of garments was rendered comprehensible by Clark's cleverly designed sets, which focused on the varied ways in which fashion uses historical references. The exhibition's theatrical staging connected to 18th-century Commedia del Arte shows and masquerades, and linked directly to

1. A tableau from the *Malign Muses* exhibition held at the Mode Museum in Antwerp in 2005, designed and curated by Judith Clark

contemporary designers' use of drama and visual excess in their seasonal catwalk shows.

Malign Muses was later staged at the Victoria and Albert Museum in London, where it was renamed as *Spectres: When Fashion Turns Back*. This new title expressed one of the contradictions at the heart of fashion. Fashion is obsessed with the new, yet it continually harks to the past. Clark deployed this central opposition to great effect, encouraging visitors to think about fashion's rich history, as well as to connect it to current issues in fashion. This was achieved through the juxtaposition of garments from different periods, which used similar techniques, design motifs, or thematic concerns. It was also the result of Clark's close collaboration with fashion historian and theorist Caroline Evans. By using Evans' important insights about fashion and history from her 2003 book *Fashion at the Edge: Spectacle, Modernity and Deathliness*, Clark revealed fashion's hidden impulses. Evans shows how influences from the past haunt fashion, as they do the wider culture. Such references can add validity to a new, radical design, and connect it to a hallowed earlier ideal. This was apparent in the fragile pleats of the Mme Grès dress included in the show, which looked to classical antiquity for inspiration. Fashion can even speak of our fears of death, in its constant search for youthfulness and the new, as evoked by Dutch duo Viktor and Rolf's all-black gothic-inspired gown.

Visitors could therefore not only see the visual and material aspects of fashion's uses of history, but through a series of playfully constructed vignettes, they were able to question the garments' deeper meanings. In a continuation of the exhibition's fairground theme, a series of carefully conceived optical illusions used mirrors to trick the viewer's eye. Dresses seemed to appear then disappear, were glimpsed through spy-holes, or were magnified or reduced in size. Thus, visitors had to engage with what they were looking at, and question what they thought they could see.

They were prompted to think about what fashion means. In contrast to clothing, which is usually defined as a more stable and functional form of dress that alters only gradually, fashion thrives on novelty and change. Its cyclical, seasonally shifting styles were evoked by Toledo's circular drawing of a never-ending parade of silhouettes, each different from the next. Fashion is often also seen as a 'value' added to clothes to make them desirable to consumers. The exhibition sets' glamour and theatricality reflected the ways that catwalk shows, advertising, and fashion photography seduce and tempt viewers by showing idealized visions of garments. Equally, fashion can be seen as homogenizing, encouraging everyone to dress in a certain way, but simultaneously about a search for individuality and expression. The contrast between couture's dictatorial approaches to fashion in the mid-20th century, embodied by outfits by Dior, for example, was contrasted with the diversity of 1990s fashions to emphasize this contradiction.

This led visitors to understand the different types of fashion that can exist at any one moment. Even in Dior's heyday, other kinds of fashionable clothing were available, whether in the form of Californian designers' simple ready-to-wear styles, or Teddy boys' confrontational fashions. Fashion can emanate from a variety of sources and can be manufactured by designers and magazines, or develop organically from street level. *Malign Muses* was therefore itself a significant moment in fashion history. It united seemingly disparate elements of past and present fashions, and presented them in such a way that visitors were entertained and enthralled by its sensual display, but led to understand that fashion is more than mere surface.

As the exhibition revealed, fashion thrives on contradiction. By some, it is seen as rarefied and elite, a luxury world of couture craftsmanship and high-end retailers. For others, it is fast and throwaway, available on every high street. It is increasingly global, with new 'fashion cities' evolving each year, yet can equally be local, a micro fashion specific to a small group. It inhabits intellectual texts and renowned museums, but can be seen in television makeover shows and dedicated websites. It is this very ambiguity that makes it fascinating, and which can also provoke hostility and disdain.

Fashions can occur in any field, from academic theory to furniture design to dance styles. However, it is generally taken, especially in its singular form, to refer to fashions in clothing, and in this *Very Short Introduction* I will explore the ways in which fashion functions, as an industry, and how it connects to wider cultural, social, and economic issues. Fashion's emergence since the 1960s as a subject of serious academic debate has prompted its analysis as image, object, and text. Since then it has been examined from a number of important perspectives. The interdisciplinary nature of its study reflects its connection to historical, social, political, and economic contexts, for example, as well as to more specific issues, including gender, sexuality, ethnicity, and class.

Roland Barthes studied fashion in relation to the interplay of imagery and text in his semiotic analyses *The Fashion System* of 1967 and *The Language of Fashion*, which collected together texts from 1956 to 1969. Since the 1970s, cultural studies has become a platform from which to explore fashion and identity: Dick Hebdige's text *Subculture: The Meaning of Style* (1979), for example, showed the ways in which street fashions evolved in relation to youth cultures. In 1985, Elizabeth Wilson's book *Adorned in Dreams: Fashion and Modernity* represented an important assertion of fashion's cultural and social importance from a feminist perspective. Art history has been a significant methodology, which enables close analysis of the ways fashion interconnects with visual culture, as epitomized in the work of Anne Hollander and Aileen Ribeiro. A museum-based approach was taken by Janet Arnold, for example, who made close studies of the cut and construction of clothing by looking at garments in museum collections. Various historical approaches have been important to examine the fashion industry's nature and relationship to specific contextual issues. This area includes Beverly Lemire's work from a business perspective, and my own work, and that of Christopher Breward, in relation to cultural history. Since the 1990s, scholars from the social sciences have become particularly interested in fashion: Daniel Miller's and Joanne Entwistle's work are important examples of this trend. Caroline Evans' impressively interdisciplinary work, which crosses between these approaches, is also very significant. Fashion's study in colleges and universities has been equally diverse. It has been focused in art schools, as the academic component of design courses, but has spread to inhabit departments from art history to anthropology, as well as specialist courses at under- and postgraduate levels.

This academic interest extends to the myriad museums that house important fashion collections including the Powerhouse Museum in Sydney, the Costume Institute at the Metropolitan Museum in New York, and the Kyoto Museum. Curatorial study of fashion has produced numerous important exhibitions and the vast numbers of visitors who attend such displays testify to the widespread interest in fashion. Importantly, exhibitions provide an easily accessible connection between curators' specialist knowledge, current academic ideas and the central core of fashion, the garments themselves, and the images that help to create our ideas of what fashion is.

A vast, international fashion industry has developed since the Renaissance. Fashion is usually thought to have started in this period, as a product of developments in trade and finance, interest in individuality brought about by Humanist thought, and shifts in class structure that made visual display desirable, and attainable by a wider range of people. Dissemination of information about fashion, through engravings, travelling pedlars, letters, and, by the later 17th century, the development of fashion magazines, made fashion increasingly visible and desirable. As the fashion system developed, it grew to comprise apprenticeships, and later college courses, to educate new designers and craftspeople, manufacturing, whether by hand or later in a factory, of textile and fashion design, retailing, and a variety of promotional industries, from advertising to styling and catwalk show production. Fashion's pace began to speed up by the later 18th century, and by the time the Industrial Revolution was at its height in the second half of the 19th century had grown to encompass a range of different types of fashion. By this point, haute couture, an elite form of fashion, with garments fitted on to individual clients, had evolved in France. Couturiers were to crystallize the notion of the designer as the creator not just of handmade clothes, but also of the idea of what was fashionable at a particular time. Important early couturiers such as Lucile explored the possibilities of fashion shows to generate more publicity for her design house by presenting her elaborate designs on professional mannequins. Lucile also saw the potential of another important strand of fashion, the growing ready-to-wear trade, which had the potential to produce a large number of clothes quickly and easily and make them available to a far wider audience. Lucile's trips to America, where she sold her designs, and even wrote popular fashion columns, underlined the interrelationship between couture styles and the development of fashionable readymade garments. Although Paris dominated ideals of high fashion, cities across the world produced their own designers and styles. By the late 20th century, fashion was truly globalized, with huge brands such as Esprit and Burberry sold across the world, and greater recognition of fashions that emanated from beyond the West.

Fashion is not merely clothes, nor is it just a collection of images. Rather, it is a vibrant form of visual and material culture that plays an important role in social and cultural life. It is a major economic force, amongst the top ten industries in developing countries. It shapes our bodies, and the way we look at other people's bodies. It can enable creative freedom to express alternative identities, or dictate what is deemed beautiful and acceptable. It raises important ethical and moral questions, and connects to fine art and popular culture. Although this *Very Short Introduction* focuses on womenswear, as the dominant field of fashion design, it also considers various examples of significant menswear. It will focus on the later stages of fashion's development, while referring to important precursors from the pre-19th-century period to show how fashion has evolved. It will consider Western fashion, as the dominant fashion industry, but equally will question this dominance and show how other fashion systems have evolved and overlapped with it. I will introduce the reader to the fashion industry's interconnected fields, show how fashion is designed, made, and sold, and examine the significant ways in which it links to our social and cultural lives.

Chapter 1
Designers

For Chanel's spring 2008 couture catwalk show, a huge replica of the label's signature cardigan jacket was placed on a revolving platform at the centre of the stage. Made from wood, but painted concrete grey, this monumental 'jacket' towered over the models, who emerged from its front opening, paraded past the audience of fashion press, buyers, and celebrities, pausing in front of its interlocked double 'C' logo, and then disappeared inside this iconic emblem of Coco Chanel's legacy. The models wore a simple palette, again reflecting the label's heritage: graphic black and white was tempered with dove greys and palest pinks. Outfits were developed from the tweed cardigan jacket that literally and metaphorically dominates Chanel, but this classic garment was made contemporary, light and feminine, shredded into wispy fronds at its hem, or fitted and sequined, worn with tiny curving skirts that drew on the organic forms of seashells for their delicate silhouettes. Both the show's staging and the clothes shown epitomized the house's origins, in their combination of Coco Chanel's love of chic skirt suits, glittering costume jewellery, and tiered evening dresses, merged with current designer Karl Lagerfeld's sharp eye for the contemporary.

2. Karl Lagerfeld's 2008 version of the classic Chanel suit

Chanel's evolution as one of the most famous and influential couture houses of the 20th century highlights many of the key elements to successful fashion design, and exposes the relationships between design, culture, commerce, and, crucially, personality. Coco Chanel's emergence in the 1910s and 1920s as a prominent figure on society and fashion pages, her mythologized rise from nightclub singer to couturier, and gossip surrounding her lovers, gave her simple, modern styles an air of excitement and intrigue. Her designs were significant in their own right, and epitomized contemporary fashions for sleek, pared-down daywear, and more feminine, dramatic eveningwear. She asserted that women should dress plainly, like their maids in little black dresses, although Claude Baillén quotes Chanel as reminding women that 'simplicity doesn't mean poverty'. Her love of mixing real and costume jewellery and her borrowings from the male wardrobe became internationally famous. Coco Chanel's biography provided the publicity and interest necessary to distinguish her house, and dramatize her as a designer and personality. Importantly, her diversification into accessories, jewellery, and perfumes, and the sale of her designs to American buyers, brought the essence of her fashions to a far wider market than could afford haute couture, and secured her financial success.

In the 1980s, fashion commentator Ernestine Carter characterized Chanel's success as founded upon 'the magic of the self'. As important as Coco Chanel's undoubted design and styling skills were, it was her ability to market an idealized vision of herself, and to embody her own perfect customer, that made the label so appealing. Chanel designed herself, and then sold this image to the world. Many others have followed her example: since the 1980s, American designer Donna Karan has successfully projected an image of herself as a busy mother and businesswoman who has designed clothes for women like herself. In contrast, Donatella Versace is always photographed in high heels and ultra-glamorous, tight-fitting clothes, her jetset lifestyle mirrored in the jewel-coloured luxury of the Versace label's designs.

Karl Lagerfeld, Chanel's present designer, represents a variation on this theme; rather than embodying the lifestyle of his customers, his personal style denotes his status as a cultured aesthete. If Coco Chanel was a fashion icon to her followers, embodying a modernist ideal of chic, streamlined femininity in the early 20th century, then Lagerfeld is a Regency dandy remodelled for contemporary times. The key elements of his personal style have remained constant throughout his stewardship of Chanel: dark suits, long hair pulled back into a ponytail and at times powdered white. Combined with the constantly flicking black fan he used to carry, his image harks back to the ancien régime. This evokes the elite status of couture, and the consistency of Chanel style, while his involvement in various art and pop cultural projects maintains his profile at the forefront of fashion.

When Chanel died in 1971, the house lost its cachet and its sales and fashion credibility dwindled. In Lagerfeld's hands it has been revitalized. Since his arrival in 1983, he has designed collections for couture, ready-to-wear, and accessories that have balanced the need for a coherent signature, and the equally important desire for fashions that reflect and anticipate what women want to wear. Lagerfeld's experience in freelancing for various ready-to-wear labels, including Chloé and Fendi, had proved his design skills and his crucial ability to create clothes that set fashions, and flatter women's bodies. He merged high and popular culture references to maintain Chanel's relevance, and to invigorate its fashion status. His spring 2008 Chanel couture collection demonstrated this and showed his business acumen. While he kept older, loyal customers in mind with his variations on the cardigan jacket, the collection's tone was youthful, with girlish flounces and froths of light fabrics counterpoised with its more sombre tones. Lagerfeld therefore looked towards the future to ensure Chanel's survival, encouraging new, younger clients to wear this iconic label.

Evolution of the couturier

Historically, most clothing was made at home, or fabrics and trimmings were bought from a range of shops and made up by local tailors and dressmakers. By the end of the 17th century, certain tailors, particularly in London's Savile Row, were establishing their names as the most accomplished and fashionable, with men travelling from other countries to have suits made for them by names such as Henry Poole. Although specific tailoring firms would be fashionable at particular times, menswear designers were not to achieve the status and kudos of their womenswear counterparts until the second half of the 20th century. The term 'tailor' evoked a collaborative practice, both in terms of the range of craftsmen involved in making suits, and the close discussions with clients that shaped the choice of fabric, style, and cut of the garments. In contrast, by the late 18th century, the creators of women's fashions had begun to evolve an individual aura. This reflected the greater scope for creativity and fantasy in womenswear. It was also dependent upon the distinct relationship that gradually developed between aristocratic fashion leaders and the people who made their clothes. While even the most noted tailors worked closely with their clients on the design of their clothes, women's dressmakers began to dictate styles.

Although fashion has remained an essentially collaborative process, in terms of the number of people involved in its production, it came to be associated with the idea of a single individual's design skills and fashion vision. The most famous early example of this shift was Rose Bertin, who created outfits and accessories for Marie Antoinette and a host of European and Russian aristocrats in the late 18th century. She was a *marchande des modes*, which meant she added trimmings to gowns. However, the *marchande des modes*' role began to change, in part as a response to Bertin's skill at creating a fashionable look. She drew inspiration from contemporary events, crafting a headdress incorporating a hot air balloon in honour of the Montgolfier brothers' balloon flights in the 1780s, for example. She generated publicity with such creations, and although other *marchandes des modes*, including Madames Eloffe and Mouillard, were also famous at this time, it was Bertin who best expressed the ebullience of contemporary Parisian fashion.

In 1776, France replaced its guild system with new corporations, and raised the status of the *marchandes des modes*, allowing them to make dresses, rather than just trim them. Bertin was the first Master of their corporation, which increased her fashion prominence. She dressed the 'grande Pandora', a doll clothed in the latest fashions, which was sent to European towns and to the American colonies. It was one of the main ways to propagate fashions before the regular publication of fashion magazines. In this way, Bertin helped to disseminate Parisian fashion, and to assert its dominance of womenswear. Her development of a wide customer base and her close relationship with the French queen ensured her fashion status. Significantly, contemporary commentators noted with horror that Bertin behaved as though she was equal to her aristocratic clients. Her elevated status was another important shift that set the stage for the dictatorial ways of many designers. She was aware of her power and confident of the importance of her work, creating fashions, but also fashioning the image of her customers, who relied on her for their own status as fashion leaders. Indeed, her boutique, the Grand Mogul in Paris, was so successful that she opened a branch in London. Her innovative styling and witty references to both historical and contemporary events showed her design skills, as well as her awareness of the importance of generating publicity. She therefore became a precursor to the couturiers, who were to evolve their own status as dictators of fashion in the 19th century.

The French Revolution effected a temporary halt in information about Parisian fashions reaching the rest of the world.

However, once this was over, the luxury trades in France were quickly re-established, and various dressmakers began to distinguish themselves as the most fashionable. Louis Hyppolite Leroy defined the fashionable style of Empress Josephine and other women of the Napoleonic court, as well as a range of European royalty. In the 1830s, names such as Victorine became well known, raising themselves above the ranks of anonymous dressmakers. Leroy and Victorine, like Bertin before them, sought to create designs and set fashions, and to assert their own prominence, as well as that of their titled clientele. However, most dressmakers, even those with aristocratic customers, did not originate designs. Instead, they provided permutations of existing styles, adapted to suit the individual customer. Styles were copied from the most famous dressmaking establishments or from fashion plates.

However, alongside leading dressmakers, there was another aspect of the fashion industry that was also involved in the evolution of the idea of the fashion designer. Art historian Françoise Tetart Vittu has shown that some artists worked in ways that mirror freelance designers today, with dressmakers buying highly detailed drawings of fashions from them. These would then be used as templates for garments, and would even be sent to customers as samples. Advertisements for the dressmakers would be attached to the back of the illustrations, along with prices for the outfit shown. By the middle of the 19th century, artists such as Charles Pilatte advertised themselves as 'fashion and costume designers' and appeared in Paris directories of the time under a list of 'industrial designers'.

The idea of clothing needing to be designed by someone with fashion authority, and with particular skills in defining a silhouette, cut, and decoration, was evolving across the Western world. Each town would have its most fashionable dressmakers, and designs themselves were gaining commercial value as fashions began to change more rapidly along with the public's desire for new styles. For the idea of the fashion designer to crystallize, there needed to be not only creative individuals ready to generate new fashions, but a growing demand for novelty and innovation. The 19th century saw the rise of the bourgeoisie and wealthy industrialists, whose newly found status was in part constructed through visual display, in their homes and, even more importantly, their clothes. Couture became a source of exclusivity and luxury for wider groups of women, with Americans amongst the most prolific customers in the second half of the century.

Added to this was the growth of fashion media, photography, and by the end of the century, film, which disseminated imagery of fashion more widely than ever before, and fuelled women's desire for more variety and quicker turnover of styles. As the huge growth in cities led to greater anonymity, fashion became a major way to formulate identity and to make social, cultural, and financial status visible. It was also a source of pleasure and sensuality, with Parisian couture at the apex of this realm of fantasy and luxury.

While 'industrial designers' supplied fashion designs to the wider dressmaking trades, it was the evolution of the couturier that was to establish the role and image of the fashion designer. Although Charles Frederick Worth, the most famous couturier of the 1850s, succeeded in part because of sound business practices, this side of his work was masked by the drama of his creations, and his persona as a creative artist whose fashion pronouncements were to be followed without question. An Englishman who had honed his skills in the dressmaking section of department stores, he was able to distinguish himself early on in his career in part because he was a man in a profession dominated by women. Indeed, in *All the Year Round* in February 1863, Charles Dickens remarked with horror at the rise of the 'bearded milliner'. As a man, Worth could promote himself in ways that would be seen as inappropriate for a woman, and he could treat his female

3. Paul Poiret's delicate Empire line gown, drawn by Georges Lepape in 1911

clients differently, irrespective of their rank. His most famous designs comprised froths of ivory tulle, creating clouds around the wearer that would glimmer in candlelit ballrooms as the beading and sequins embroidered between the layers caught the light.

Other couturiers were also rising to prominence, often propelled to fame by their royal customers. In England, John Redfern responded to the changing role of women in the period by producing couture gowns based on men's suits, and sporty ensembles for yachting. In France, female couturiers such as Jeanne Paquin made garments that shaped women's bodies and epitomized the ideal of the Parisienne. Many customers came from America, as Paris continued to lead fashion. Fashion houses, partly to raise the status of the designer, and partly to provide a recognizable identity and personality to promote each label, asserted the idea of the couturier as an innovator and artist. Cecil Beaton described women in the Edwardian period who tried to keep the names of their dressmakers secret. Such women wanted to be credited for their own fashion sense and remain better known than their couturier. However, couture houses were already evolving their own recognizable styles, which conferred fashion status on the women who wore them.

In the first decades of the 20th century, designers such as Paul Poiret and Lucile became internationally famous. They dressed theatrical stars, aristocrats and the wealthy, and promoted their own identities as decadent socialites in their own right. Poiret was a fashion designer in the modern sense of the phrase. He was known for his signature luxurious style, and the radical, seasonally changing silhouettes he created. Georges Lepape's fashion illustration shows Poiret's famous Empire line silhouette of 1911, which broke away from the tightly corseted fashions of the Edwardian period. His lavishly embroidered gowns and opera coats were inspired by contemporary art and design, from modernism to the Ballets Russes, and the aura of his potent couture image was disseminated still further by sales of his own perfume line. Poiret's contemporaries were equally adept at harnessing modern advertising and marketing methods to create the image of their fashion house. Most sold their designs to American wholesalers, for them to make up a strictly defined number of each model they had bought. This generated income for the couture houses, alongside money from the individually made garments that were the very definition of haute couture.

The interwar period was a high point for couture, when Madeleine Vionnet, Elsa Schiaparelli, Coco Chanel, and others defined the idea of modern femininity through their creations. Their success underlined the fact that fashion has long been one of the few arenas in which women could be successful as creators and entrepreneurs, heading their own businesses and providing work for countless other women in their couture studios. Indeed, couture is a collaborative venture, with big fashion houses comprising numerous studios each working on a different aspect of a design, for example tailoring or draping or decoration, including beading or feathers. Despite the number of people involved in the creation of each garment, the idea of the fashion designer has evolved in line with the idea of the artist as a creative individual. This is partly because design and innovation are the most valued aspects of fashion, since they are the basis for each collection and viewed as the most creative element of the process. Importantly, this focus on the individual is also a successful promotional tool, as it gives a focus for the identity of a fashion label, and quite literally, provides a 'face' for the design house.

Although not governed by the strict rules that apply to Parisian haute couture, other countries have developed their own couturiers and made-to-order industries. For example, in 1930s London, Norman Hartnell and Victor Stiebel asserted themselves as fashion designers rather than just court dressmakers, while in New York, Valentina evolved a dramatically simple style that drew on contemporary dance to create an American fashion identity, and in the 1960s in Rome, Valentino promoted a distinctively Italian form of couture that relied on overtly feminine luxury.

In the post-war period, fabric and labour costs increased, making couture even more expensive. Designers such as Christian Dior revelled in excess, after the hardships of the 1940s, with their focus on the traditions of couture craftsmanship, and led a decade in which couture continued to dominate international fashion trends. Since the 1960s, despite the rise of throwaway youth fashions and the global fame of ready-to-wear designers, couture has maintained its visibility. Its significance has shifted, but certain couturiers, such as John Galliano at Dior, Alber Elbaz at Lanvin, and Lagerfeld at Chanel, are still able to set fashions that disseminate through all levels of the market. Despite a falling number of clients, ready-to-wear lines, accessories, perfumes, and a huge number of other licensed ranges place couture at the forefront of the huge global luxury market. Although there are fewer haute couture customers in Europe, other markets have periodically emerged. Oil wealth increased sales in the Middle East in the 1980s, as did the strong dollar and love of display in Reagan's America, while the enormous wealth generated in post-Communist Russia has provided more clients in the early 21st century. Combined with the prominence of celebrity culture and the rise of the red carpet dress, couturiers continue to produce seasonal collections. Even if these one-off designs do not make a profit themselves, the huge quantity of publicity they generate asserts the continued importance of the designer at the heart of the couture industry.

Evolution of the ready-to-wear designer

In her 1937 book *Clothes Line*, the British fashion journalist Alison Settle wrote that the interconnected nature of the Parisian haute couture industry was crucial to its success. Fabric, dress, and accessory designers and makers were in close contact with each other, and could respond to developments within each field.

Trends were therefore identified quickly and integrated into couturiers' collections, allowing Paris to maintain its position at the forefront of fashion. Settle was also impressed by how embedded fashion was within French culture, with people of all social classes interested in clothing and style. As Settle noted, couturiers 'forecast fashion by observing life', and this approach was particularly significant in the evolution of the ready-to-wear fashion designer. Couturiers realized that many women wanted to buy clothes that were not just in line with contemporary styles, but which were made by a fashionable name.

From the early 1930s, designers began to create less expensive collections, which could reach out to this wider audience. Lucien Lelong, for example, started his 'Lelong Édition' line, selling readymade dresses at a fraction of the cost of his couture collection. Couturiers continued to work on readymade clothes; for example, in the 1950s, Jacques Fath designed a successful line for American manufacturer Joseph Halpert. However, when Pierre Cardin launched a ready-to-wear collection at Parisian department store Printemps in 1959, he was briefly expelled from the Chambre Syndicale de la Haute Couture, which regulates the couture industry, for branching out in this way without seeking permission. At the same time, Cardin was exploring the potential market in the Far East, in his quest for global success. These moves, when considered in relation to his bold, modern style, were part of a shift in emphasis in French fashion, as couturiers strove to maintain their influence in response to the increasing success of ready-to-wear designers. In 1966, the launch of Yves Saint Laurent's Rive Gauche boutiques chimed with popular culture and recognized women's changing roles with trouser suits and vividly coloured separates. Saint Laurent showed that couturiers could set fashions through their ready-to-wear collections too. In a 1994 interview with Alison Rawsthorn, one customer, Susan Train, described his new line as 'so exciting. You could buy an entire wardrobe there: everything you needed.' However, the 1960s is generally viewed as a key moment when mass-produced, youthful ready-to-wear began to lead fashion in a way it never had before. American designers such as Bonnie Cashin, British names, for example Mary Quant, and Italians, including Pucci, were all asserting their fashion influence at different levels of the market and shaping the way fashion was designed, sold, and worn.

While ready-to-wear clothes had been developing independently of Parisian haute couture since the 17th century, it was not until the 1920s that they were designed and marketed principally on their fashion values, rather than their price or quality. In Paris, this meant couturiers spent the following decades making agreements with department stores internationally to sell versions of their couture garments, as well as evolving their own lines. In America, manufacturers, including Townley, and stores such as Saks Fifth Avenue were quick to employ designers to work anonymously to develop fashion lines.

It was in the 1930s that these designers began to emerge from anonymous back rooms and have their names included on labels. In New York, Dorothy Shaver, vice-president of specialty store Lord & Taylor, began a series of campaigns promoting American ready-to-wear and made-to-order designers alongside each other. Window and in-store displays included photographs of named designers shown with their fashion collections, encouraging a cult of personality that had previously been reserved for couturiers. This was partly an attempt to encourage homegrown talent while the hardships of the Great Depression made trips to Paris to source fashions too costly. It was also symptomatic of fashion designers' need to group together in order to promote the status of their own fashion capitals. While Paris maintained its place at the heart of fashion, by the 1940s, in the absence of French influence during the war, New York had begun to assert its fashion status. Subsequently, cities across the world have followed the same process, investing in design education, holding their own fashion weeks to promote their designers' collections, and seeking to sell both domestically and internationally. The role of the fashion designer is vital to this process, once again providing creative impetus combined with recognizable faces that could be used as the basis for promotional campaigns. In the 1980s, Antwerp and Tokyo each demonstrated their ability to develop distinctive fashion designers, with the rise of names such as Ann Demeulemeester and Dries Van Noten in Belgium, and Rei Kawakubo of Comme des Garçons and Yohji Yamamoto in Japan. By the early 21st century, China and India, amongst others, were also investing in their fashion industries and cultivating their own seasonal shows.

The way designers are trained influences their approach to creating a collection. For example, British art colleges emphasize the importance of research and individual creativity. This stress upon the artistic elements of the creative process produces designers, such as Alexander McQueen, who are inspired by history, fine art, and film. His collections have been staged on themed sets, with models writhing in a huge glass box, or sprayed by a mechanical paint jet as they turn slowly on a rotating platform. His models are styled as characters, part of a narrative that is told through clothes and setting. His cinematic approach was apparent in his spring 2008 collection, which was inspired by the 1968 film *They Shoot Horses Don't They?* This prompted a Depression-era dance marathon theme, choreographed by avant-garde dancer Michael Clark. Models slid across a dance floor, dressed in fluid tea dresses and worn denims, their skin glistening and eyes glazed as if they had been dancing for hours, half-carried, half-dragged by male dancers. McQueen's promotion of fashion as spectacle underpins the success of his label and testifies to his creative appeal.

In contrast, colleges in the United States tend to encourage designers to focus on creating clothes for a particular customer group and to keep business considerations and ease of manufacture at the forefront of their minds. They use industrial design as a model to promote an ideal of democratic design that aims towards the greatest number of potential consumers. The work of designers such as Bonnie Cashin from the 1930s to 1980s is a good example of how this approach can lead to measured collections that aim to address women's clothing needs. Her designs looked streamlined, while demonstrating close attention to detail, with interesting buttons or belt buckles to enliven their plain silhouettes. In 1956, Cashin told writer Beryl Williams that she believed that 75% of a woman's wardrobe comprised 'timeless' pieces, and stated that 'all those clothes of mine were perfectly simple... they were simply the kind of clothes I liked to wear myself'. She designed lifestyle clothes for work, socializing, and leisure time, while promoting herself as the embodiment of her easy-to-wear styles. This type of design has come to characterize American fashion, but its simplicity can make it difficult to define a distinct image for a label. Between the late 1970s and late 1990s, Calvin Klein used controversial advertising campaigns to gain publicity for his clothing and perfume lines. Imagery such as the photograph of a teenage Kate Moss, nude and androgynous for Obsession in 1992, provided him with an edgy, contemporary image that belied the conservative styling of many of his designs.

While these designers have relied on the idea of the individual as fashion originator, many fashion houses employ whole teams of designers to produce their lines. For this reason, Belgian designer Martin Margiela refuses to give individual interviews, and avoids having his photograph taken. All correspondence and press releases are signed 'Maison Martin Margiela'. In 2001, in a faxed interview with fashion journalist Susannah Frankel on Maison Margiela's alternative approach to fashion, the choice to use non-professional models was explained as part of this overall strategy: 'We have nothing against professional or "top models" as individuals at all, we just feel that we prefer to focus on the clothes and not all that is put around them in and by the media.' His labels are blank or stamped with the number of the collection a garment comes from. This deflects attention from the individual designer and suggests the collaborations necessary to make a fashion collection, while acting to distinguish his work. For other designers, the emphasis is placed more on their celebrity customers, who add a glamorous aura to their collections. In the early 21st century, American designer Zac Posen benefited from young Hollywood stars, including Natalie Portman, wearing his dresses on the red carpet. The coverage that stars receive at such events can boost sales for new designers, as well as established fashion houses, as shown by Julianne Moore's successful championing of Stefano Pilati's designs for Yves Saint Laurent.

Menswear designers have also risen to the fore during the 20th century, although they do not command the same level of attention as womenswear designers. Designs tend to focus on suiting or leisurewear, and menswear is perceived as lacking the spectacle and excitement attached to womenswear. However, designer names began to emerge in the 1960s, with, for example, Mr Fish in London and Nino Cerruti in Italy. Both exploited the more flamboyant designs of the decade to the full, with vibrant colours and pattern and unisex elements included in their designs. Michael Fish evolved his style while working within the elite environment of Savile Row, before opening his own boutique in 1966. Meanwhile, Cerruti's sleek designs evolved out of his family's fabric business, launching his first full menswear collection in 1967. Parisian couturiers also branched out into menswear design, including Yves Saint Laurent in 1974. In the 1980s, designers continued to explore the parameters of menswear design, focusing on adaptations of the traditional suit. Giorgio Armani stripped out its stiff underpinnings to create soft, unstructured jackets in wools and linens, while Vivienne Westwood tested the limits of gender boundaries in fashion, adding beading and embroidery to jackets or putting male models in skirts and leggings.

Since the 1990s, the rich colours and textures of Dries Van Noten's collections, and the innovative fabrics in Prada's designs, for

4. Hedi Slimane's highly influential skinny silhouette for spring 2005

example, have shown that menswear design can attract attention for subtle details. The growth of male grooming and fitness culture has added to interest in the field. In the early 21st century, designers such as Raf Simons, and especially Hedi Slimane, who designed for Dior Homme from 2000 to 2007, developed a skinny silhouette for men, which was very influential. Slimane's narrow trousers, monochrome palette, and tightly fitted jackets required a youthful physique that was androgynous and uncompromising. The speed with which celebrities and rock stars, as well as high street stores, adopted this look demonstrated the power and influence that confident menswear design could have.

One of the strongest reference points in menswear collections since the 1960s has been subcultural style. From the narrow suits worn by sixties Mods to the pastel leisurewear of eighties Casuals, street style balances individuality and group identity. It therefore appeals to many men's search for clothing that acts as a kind of uniform, while simultaneously allowing them to add their own personal touches. Members of subcultures in many ways design themselves through their style, by customizing garments or breaking mainstream rules about how clothing should be worn or combined. In the late 1970s, this DIY ethos was epitomized by Punks, who adorned their clothes with slogans and safety pins, ripping the fabric and creating their own individual interpretations of classic leather biker jackets and T-shirts. While since the mid-1990s Japanese teenagers of both sexes have made their own clothes, combining them with elements of traditional dress such as obi sashes to create a wide variety of styles, united by their love of exaggeration and fantasy. By referring to these practices, fashion designers can add a seemingly rebellious edge to their collections.

Indeed, since the 1990s, fashion consumers have increasingly sought to individualize their look by customizing garments and mixing designer, high street, and vintage

5. Japanese street fashion brings together references to East and West, old and new

clothes. This enables them to act as designers themselves, if not always of individual garments, then of the look and image they wish to convey. The idea of the 'fashion victim' of the 1980s who wore complete outfits by one designer has led many wearers in reaction to seek to express their own creativity through the way they adapt and style themselves, rather than relying on designers to construct an image for them. This approach mimics both subcultural style and the work of professional stylists. It reflects a developing knowingness amongst certain consumers, and their wish to be both part of fashion yet above its dictates. While the 20th century undoubtedly saw the establishment of the designer name as the guiding force in fashion, this has not gone unchallenged. The 1980s was perhaps the apex of the cult of the designer, and while many labels are still revered, they must now compete both with a wider number of global rivals and with many consumers' desire to design themselves, rather than unquestioningly obey fashion trends.

Chapter 2
Art

Andy Warhol's *Diamond Dust Shoes* of 1981 shows a cluttered array of bright, jewel-coloured women's pumps set against an inky-black background. Based on a photographic screen print, the shoes are shot from above, the viewer seemingly looking down on a wardrobe floor, crowded with odd shoes. A vertiginous tangerine stiletto presses up next to a more demure, tomato-red rounded toe, while a brocaded midnight-blue evening slipper lies next to a salmon-pink, bow-adorned court shoe. The colours are overlaid onto the image and produce a cartoonish pastiche of the multitude of styles and shapes of shoes available.

The picture is cropped to give the impression that the pile of shoes is limitless, glimpses of the pointed tip of a lilac boot, for example, peek in at the edge of the frame. The image is carefully composed; despite the apparent jumble, each shoe is artfully displayed, with just enough inner labels visible to reinforce their high fashion status. It evokes the fashion image and the shoe shop, and thus refers to the combination of visual and literal consumption so fundamental to fashion. Warhol's painting is slick with the shine of polymer paint, an effect enhanced by the fact that the whole surface of the image is scattered with 'diamond dust', which glitters and dazzles the viewer as it catches

6. Andy Warhol's *Diamond Dust Shoes* of 1981 shows the glamour and seduction of footwear design

the light. Its shimmering surface makes explicit reference to fashion's glamour and ability to transform the mundane.

In the late 1950s, Warhol had worked as a commercial artist, with clients including I. Miller shoes. His drawings for them were sinuous and light, graphically evoking shoes' seductive appeal. His alliance to commerce and love of popular culture meant that fashion was a perfect subject for him. It featured in his screen prints and other artworks, and he continually used clothing and accessories, including his famous silver wigs, to alter and play with his own identity. In the 1960s, he opened a boutique, Paraphernalia, selling a mix of fashionable labels such as Betsey Johnson and Foale and Tuffin. Paraphernalia's launch included a performance by the Velvet Underground, and therefore united the varied strands of Warhol's entrepreneurial artworks. He understood the alliance between fashion, art, music, and popular culture that was crystallized during this decade. The marriage of avant-garde pop music with throwaway, experimental clothes that relied on brightly coloured metals, plastics, and clashing prints did not merely express the creative excitement of the period, it helped to define its parameters. For Warhol, there was no hierarchy of art or design forms. Fashion was not condemned for its commercial imperative, or its transience. Instead, these inherent qualities were flaunted in his work, as part of his fascination with the fast pace of contemporary life. Thus, the dazzling surface of *Diamond Dust Shoes* celebrated fashion's focus on outer appearance and spectacle, while his boutique brought attention to the commercial transactions and consumerist drive at the heart of fashion, and indeed much of the contemporary art market. In Warhol's art, fashion's supposed flaws of ephemerality and materialism become comments on the culture that spawned it. For Warhol, elements of mass culture and high-end luxury could coexist, in the same way that they did in fashion magazines or Hollywood films. In his work, multiples and one-offs were given equal status, and he moved easily from one medium to another, fascinated as much by the possibilities of film as of screen printing or graphic design. Rather than feeling this limited his work, or that commerce should be excluded from art for it to be legitimate, Warhol embraced contradictions. In his 1977 book *The Philosophy of Andy Warhol (From A to B and Back Again)*, he wrote of the blurred boundaries that drove his art:

> Business art is the step that comes after Art. I started as a commercial artist, and now I want to finish as a business artist. After I did the thing called 'art' or whatever it's called, I went into business art. I wanted to be an Art Businessman or a Business Artist. Being good in business is the most fascinating kind of art.

Since the mid-19th century, fashion had increased in pace, reached out to a wider audience, embraced industrial processes, and used spectacular methods to sell its wares. Art also went through this cycle of change; art markets grew to embrace the middle classes, mechanical reproduction altered ideas of exclusivity, and institutional and private galleries re-thought the way artworks were displayed and sold. There also existed a crossover in thematic concerns between the two disciplines, from issues of identity and morality, to concerns over the way the artist or designer was perceived within the wider culture, and a focus on representation of and play with the body.

Fashion is occasionally cast as art, but this is problematic. Some designers have appropriated aspects of art practice in their own work, but they remain within the structure of the fashion industry and use these borrowed methods to explore the nature of fashion itself. When, for example, in their early career, Viktor and Rolf decided just to stage fashion shows rather than produce any saleable clothes, their designs became one-offs, rare pieces that existed only as comments on the role of the show within the fashion system, rather than wearable garments. However, their work remained within the context of the fashion world, discussed and reviewed by fashion journalists. It seemed like evolving advertising campaigns for the collections they later showed, which were put into production. Their work also served to underline the differences between types of designers. Viktor and Rolf's interpretation of fashion incorporated a fascination with the role of the show, and its potential to test the boundaries of spectacle and display. They slip between art, theatre, and film in the staging of their collections. For autumn/winter 2000, the designers slowly dressed a single model in layer upon layer of garments, until she wore the whole collection. This commented on the process of fitting clothes on the body, which lies at the core of traditional fashion design. The exaggerated scale of the final clothes she was swathed in seemed to turn her into an immobile doll, a living mannequin, and the plaything of the designers. In 2002/3's show, all the clothes were bright cobalt, and acted like the blue screen used to shoot special effects in television and cinema. Film was projected across the models' bodies, which made their figures disappear and seem to flicker as images hovered across their surface.

In Viktor and Rolf's designs and presentations, artistic methods are used to comment on the practice of fashion, but this does not necessarily turn their fashion into art. Their work is shown in the context of the international fashion weeks, it is directed to a fashion audience, and addresses the way clothing and body interact. Even when they were not putting their clothing into production, they followed the fashion seasons, and importantly, they adhered to the fundamental elements of fashion: fabric and body.

Fashion is sometimes compared to art in order to give it greater validity, depth, and purpose. However, this perhaps reveals more about Western concern that fashion lacks these qualities than it does about fashion's actual significance. A Balenciaga dress from the 1950s, when displayed in a pristine glass case in a gallery, may appear like a work of art. However, it does not need to be described as such in order to convey its value or the skill that went into its creation. Like other design forms, such as architecture, fashion has its own particular concerns that prevent it from ever being purely art, craft, or industrial design. It is, rather, a three-dimensional design form that incorporates elements of all these approaches. It is Balenciaga's exacting eye for precise form that brings balance and drama to the drape and structure of the fabric, combined with the craft skill of his atelier workers, that turns it into an exceptional piece of fashion clothing. It does not need to be called art in order to validate its status, and this term ignores the reason, beyond his desire to create and test the parameters of fashion design, that Balenciaga's dresses were brought into being: to clothe a woman, and, ultimately, to sell more designs. This should not be seen to diminish his achievement, but to help to understand the way he has worked to exploit these 'limitations' to create fashions that can inspire the viewer as much as the wearer.

Fashion should be understood on its own terms, and this makes its interactions with other aspects of art and culture more interesting. It opens up the way art, design, and commerce connect and overlap in some practitioners' work. Indeed, one of the things that makes fashion so fascinating, and for some, so problematic, is the fact that it continually appropriates, reconfigures, and tests the boundaries of these definitions. Thus, fashion can highlight tensions concerning what is valued in a culture. Designers and artists as diverse as Andy Warhol and Viktor and Rolf produced work that played upon cultural contradictions and attitudes. In fashion's case, focus on body and cloth, and the fact that it is, usually, designed to be worn and sold, distinguishes it from fine art. However, this does not prevent fashion from being meaningful, and the art world's continued fascination with fashion underlines its cultural significance.

Portraiture and identity

Perhaps the most obvious connection between fashion and art is the role clothing has played within portrait painting. In the 16th century, the Reformation's impact in Northern Europe led to a decline in commissions for religious paintings, and artists therefore turned to other subject matter. Since the Renaissance, humanist interest in the individual added to many members of the nobility's desire to be portrayed by artists. The growth of portraiture established a relationship between artist and sitter, and between fashion and representation. Holbein's paintings of the royal court and nobility of Northern Europe explored the visual effects that can be conveyed in paint, and suggested the tactile differences between, for example, satin, velvet, and wool. Holbein's precision is apparent in the detailed drawings that he undertook in preparation for his portraits. Jewellery was sketched in all its intricacy, and the delicate layers of muslin, linen, and stiffening that women's headdresses comprised were explored with as much care as sitters' faces and expressions. Holbein understood the role fashionable dress played in conveying his clients' wealth and power, as well as their gender and status. These attributes were made manifest in his paintings, and turned into mementoes not just of past clothing styles, but of fashion's role in constructing an identity that could be read and understood by contemporaries. His portraits of Henry VIII portray the period's visual excess, with padded layers of silk and brocade to add size and grandeur to his figure. Gold and jewelled trimmings and accessories increased this effect, and fabrics were slashed to reveal further lavish garments beneath. His portraits of women were equally rich in detail. Even his sombre 1538 painting of Christina of Denmark wearing mourning dress revealed the fabric's richness. The soft shine of her long black satin gown is emphasized by the light falling on its deep folds and full, gathered shoulders. This is contrasted with the tawny red-brown fur that lines the gown, and the supple pale leather of her gloves. Holbein's compositions, like those of artists across Europe at the time, placed focus on the sitters' faces, while also giving great emphasis to displaying their clothing's splendour.

This spectacle of fabric and jewellery is present in the work of artists from Titian to Hilliard. Even when, as in the portrait of Christina of Denmark, the dress is restrained and undecorated, the lushness of the materials plays a major role in establishing the sitter's status. The significance of this display would have been easily comprehensible to contemporaries. Textiles were hugely expensive, and therefore greatly valued. The ability to purchase and wear an array of cloth of gold and silk velvets asserted the sitter's wealth. Glimpses of white shirts and smocks, worn beneath the layers of outer garments, further reinforced sitters' standing. Cleanliness was a mark of status, and servants were needed to keep linens laundered and white, and ruffs starched and properly pressed into their complicated shape.

Art did not merely serve to advertise royal and noble status; it also displayed character, taste, and the sitter's relationship to fashion. While artists such as Holbein strove to paint contemporary fashions accurately, as part of the overall realist approach of his work, others used greater artistic licence. During the 17th century, Van Dyck and others often showed sitters in draped fabrics that curved around the body in impossible ways. They framed the body in allegorical dress, intended to evoke Greek muses or goddesses. Women were swathed in pastel satins that seemed to fly around the body and float over the surface of the skin. Men were shown in outfits that were part reality, part fancy dress. While Van Dyck also painted fashionable dress, he frequently imposed his own unifying taste for light-reflecting surfaces and uninterrupted planes of colour. Thus, art mediated fashion, it was not just a record of what was worn and how, but of ideals of beauty, luxury, and taste.

Art's relationship to fashion became more complex as fashions began to change seasonally during the 18th century, and some artists became uneasy about the effect of this on the status of their work. Some portraitists, such as Joshua Reynolds, wanted to strive for longevity and create a painting that would transcend its time. Fashion seemed to hamper these ambitions; it pulled a painting back into the time when it was created. As styles changed yearly, if not seasonally, portraits were precisely datable. While for Van Dyck and his sitters classicized clothing was part of a playful interest in fancy dress, for Reynolds it was a serious attempt to break from fashion and propose an alternative way to guarantee the relevance of portraits for posterity. He therefore strove to erase fashion from his art, painting sitters in imagined swathes of fabric to relate the figure to classical drapery seen in ancient statuary. Fashion's power to shape how body and beauty are perceived disrupted Reynolds' intentions. Although the dress he often painted was plain, so was much of fashion in the last quarter of the 18th century, as was the long, narrow silhouette that he favoured. The sitter's desire to be seen as modish also hampered his classicizing eye. Female clients persisted in wearing towering, powdered wigs, often topped with plumes of feathers. Their faces were also powdered white, with cheeks fashionably pinked.

This combination of the sitter's wish to be seen as fashionable and the artist's difficulty in breaking away from the dominant visual ideal of the day meant that it was almost impossible to paint a portrait that did not betray its date. In her book *Seeing through Clothes*, Anne Hollander proposed that:

> in civilised Western life the clothed figure looks more persuasive and comprehensible in art than it does in reality. Since this is so, the way clothes strike the eye comes to be mediated by current visual assumptions made in pictures of dressed people.

Hollander contends that it is not just the clothed body that is 'learnt' through its representation in art. She also argued that artists' vision is trained by contemporary fashion, and that even when a nude body is painted, the shape of the body and the way it is presented is tempered by prevailing fashionable ideals. The small, high breasts and low stomachs of Cranach's nudes of the 15th century, Rubens' full-bodied *Three Graces* of the 1630s, and Goya's clothed and nude *Maja* of the early 19th century all bear witness to the impact of the fashionable silhouette on the way the body is portrayed. In each case, the shape of the clothed body, re-formed by corsetry, padding, and overgarments, is imposed on the naked figure. Thus, the relationship between portraiture and fashion is deeply embedded, and demonstrates the interconnected nature of visual culture at any given time.

This interrelationship was to become more explicit in the 19th century, with artists such as Cézanne, Degas, and Monet using fashion plates as templates for their female figures and the clothes they wore. Since many people see fashion through imagery, whether paintings, drawings, fashion plates, or later photographs, the viewer, like the artist, is coached to understand the clothed bodies she sees around her in terms of these representations. Indeed, Aileen Ribeiro has taken this idea further to suggest the materialism involved in commissioning and purchasing art was part of the same consumer culture that saw the growth of the fashion industry in the second half of the 19th century, and the comparably huge amounts charged by leading portraitists and couturiers such as Charles Frederick Worth. Ribeiro cites as evidence of this close alliance Margaret Oliphant's observation in her book *Dress* of 1878 that 'there is now a class who dress after pictures and when they buy a gown ask "will it paint?"'.

Perhaps the most compelling example of this blurred line between fashion and its representation is the collection of over four hundred photographs taken by Pierre-Louis Pierson between 1856 and 1895 of Virginia Verasis, the Comtesse de Castiglione. She took an active role in the way she was dressed, styled, and posed. She therefore took on the role of artist herself, controlling both her presentation through fashion and her representation in the photographs. Her elaborately decorated dresses of the mid-19th century act like fashion photographs, while going beyond the remit of fashion imagery to construct an individual's relationship to dress. Castiglione was aware that she was giving a performance in each image, and staged herself within a suitable environment, whether a studio setting or on a balcony. She demonstrated the power of 'self-fashioning', using dress to define and construct the way she was perceived and her body displayed. For her, the interconnections between fashion and art were a powerful tool to

7. Countess Castiglione created her own image in numerous photographs of the mid-19th century

allow experimentation with various identities, since, as Pierre Apraxine and Xavier Demarge have argued:

> Castiglione's use of her own body – the primary source of her art – and the way in which she orchestrated her public appearances [presaged]... such contemporary developments as body art and performance art.

Fashion's significant role in visual culture, and the inextricable link between actual garments and their representation in art and magazines, meant artists tended to be ambiguous about its power. While portraitists including Winterhalter and John Singer Sargent used their sitters' fashionable dress to shape compositions and suggest the status and character of their sitters, others, most notably the Pre-Raphaelites, rejected fashion's pervasive hold on ideals of beauty, style, and taste. By the 1870s, an Aesthetic dress movement had emerged which sought to offer an alternative to fashion's restrictive definitions of the body, in particular the role of corsetry in moulding women's bodies. Men and women turned instead to looser-fitting historicized styles. However, Aesthetic dress itself became a fashion, although it crystallized the idea that artists and those interested in fine art might dress in an alternative, anti-fashion style. While they might refuse contemporary trends, their studied indifference to their dress is implicit recognition of fashion's role in shaping how they are perceived, and the power of dress in fashioning identity.

Collaborations and representations

During the 20th century, there were numerous cross-fertilizations and collaborations between art and fashion. Haute couture's developing aesthetic sensibilities combined advanced craft skills with individual designers' vision and pressure to create a strong business practice to ensure prolonged success. Couturiers sought to establish their design houses' identities in relation to contemporary beauty ideals and this necessarily saw them look to modern art as a visual prompt and inspiration. In Paul Poiret's hands, this meant an exploration of notions of the exotic, and, like Matisse, he travelled to Morocco to find alternatives to Western approaches to colour and form. Poiret's fantasy of rich planes of colour, draped harem trousers, and loose tunics contributed to an ideal of femininity that had been increasingly apparent in both popular and elite culture since the later 19th century. Poiret and his wife Denise were photographed in orientalized robes, reclining on sofas at their infamous 'One Thousand and Second Night' party. When viewed in conjunction with Poiret's designs, these images promoted his couture house as luxurious and decadent. Importantly, they also positioned him as uncompromisingly modern, despite the historical references that underpinned many of his garments. Poiret was aware that he needed to cultivate an image of himself that drew on notions of the artist as an individual creative force, while also producing designs that could successfully be sold abroad, particularly to America. His work, in common with other couturiers, had to balance between the demands of the one-off outfit for a single client, which had more in common with the authenticity of fine art, and the commercial imperatives of creating designs that could be sold to and copied by manufacturers internationally. Although Poiret strove to maintain an artistic image, and drew upon such influences as the Ballets Russes, he also undertook promotional tours to Czechoslovakia and America to increase awareness of his designs amongst a wider audience.

Nancy Troy has written of this delicate relationship between fine art practice and haute couture in the first decades of the 20th century. She identified shifts in each discipline that were a response to the increasingly blurred line between popular and elite culture, and therefore also to distinctions between the 'authentic' original and the reproduction. As she noted, designers and artists tried 'to explore, control, and channel (though not necessarily to stave off) the supposedly corrupting influence of commerce and commodity culture'.

Couturiers had varied approaches both to managing these issues and to incorporating influences from contemporary art into their designs. Poiret's work flourished under the influence of vibrant, often clashing colour and emphasis on theatrical self-presentation. It is therefore unsurprising that when he made direct collaborations with artists, it was in textile designs by Matisse and Dufy, for example. Such connections between leading avant-garde artists and their equivalents within fashion seem both natural and mutually beneficial. Each side was able to experiment, exploring new ways to think about and present their ideas. Each potentially benefited by the association with another form of cutting-edge contemporary culture to combine the visual with the material. Elsa Schiaparelli staged more extensive collaborations, most famously through her work with Salvador Dali and Jean Cocteau. These connections produced clothes which gave life to Surrealist tenets, including Dali's 'lobster' decorated dress. This brought the movement's love of juxtapositions and complex relationship to notions of femininity into the physical realm, with Schiaparelli's wearers turning their bodies into statements on art, culture, and sexuality.

For Madeleine Vionnet, an interest in contemporary art's preoccupations was seen in her technical explorations of the three-dimensional planes of a garment, inspired by the fragmented representational style of Italian Futurism. Her work with Ernesto Thayaht showed a dynamic union between his spatial experiments and her concern for the relationship between body and fabric. His fashion plates of her designs made this link explicit, rendering her designs as Futurist ideals of femininity. The models' bodies and clothes were fractured to show not just their three dimensions, but to suggest their lines of movement and their intrinsic modernity.

If Poiret's association with art was through his desire for luxury and freedom in design expression, then Vionnet's was part of a search for new methods to address the body and the way it was represented. Both couturiers were also widely copied, despite the intricacy of their designs. Their concern about manufacturers' profligate use of their work exposed the contradictions inherent within modern fashion (and, indeed, art). As Troy has shown, what was at stake was not just ideals of artistic integrity; copying could also undermine their businesses and jeopardize their profits.

Given art and fashion's growing push into the commercial world, it was inevitable that artists and designers would look to mass-produced ready-to-wear as another site for collaboration. Such projects brought tensions between the two disciplines, and their relationship to industry and finance to the fore. This could be through political belief in the power of art to change the lives of the masses, as seen in Russian Constructivist Vavara Stepanova's designs of the 1920s. While most of her contemporaries shunned fashion for its ephemerality, she felt that, despite its problematic associations with capitalism and business, it was bound to become more rational, in the same way that she perceived 'daily life' in the Soviet Union to be. She therefore broke with her fellow Constructivists to state that:

> It would be a mistake to think that fashion could be eliminated or that it is an unnecessary profit-making adjunct. Fashion presents, in a readily understandable way, the complex set of lines and forms predominant in a particular time period – the external attributes of the epoch.

Fashion's ability to connect more directly with the wider community has made it an ideal medium for artists who want to connect their work to the popular sphere. This might follow in the traditions established by Poiret at the start of the 20th century, as seen in the witty prints designed by Picasso, amongst others, for a series of American textile designs in the 1950s, and used by designers including Claire McCardell. In the early 1980s, Vivienne Westwood's work with graffiti artist Keith Haring was closer in spirit to Schiaparelli's collaborations with artists. In each case, their joint work represented a common interest and intent, in Westwood and Haring's example in street culture and challenging accepted ideas of the body, which translated into clothes decorated with an artist's drawings.

The commercial and consumer ethic at the heart of many collaborations between fashion and art became more manifest in the later 20th and early 21st centuries. While Rei Kawakubo's rigorous intellectual approach to fashion is without question, it is interesting to see how successfully she has negotiated the potentially fraught relationships between artistic endeavour, fashion, and consumption. Peter Wollen has compared Japanese designers' approach to these interconnections to that of Wiener Werkstätte artists, who sought to design clothes as part of a 'total environment'. This environment includes, perhaps most significantly, the retail space, which for Comme des Garçons has become a temple for Kawakubo's design aesthetic and a site of continuing collaborations. Leading architects, including Future Systems, designed boutiques for her in New York, Tokyo, and Paris. Her interior displays ape iconic modernist works, such as the apparently haphazard design of her Warsaw guerrilla store, which made reference to Bauhaus designer Herbert Bayer's ground-breaking presentation of identical chairs fixed to the walls in the German section of the Society of French Interior Designers annual exhibition of 1930.

Kawakubo, like other designers including Agnès B, has taken this ambiguity between commercial retail space and gallery further, to hold exhibitions in her boutiques. In Comme des Garçons' Tokyo store, displays have included Cindy Sherman's photography, which itself appropriates fashion practices. Such exhibitions are not new, for example New York department store Lord & Taylor held a show on Art Deco in the late 1920s, while Selfridges in London showed Henry Moore's work in the 1930s. However, by the end of the century connections were more complex and links between the two areas more firmly embedded, especially within the work of artists and designers dealing with the body and identity.

At the start of the 21st century, the relationship between art and fashion remains as fraught as it is revealing of cultural values and subconscious desires. The lines between fashion in art and art in fashion became hazier, but so did the distinctions between the spaces in which each was shown. Shops, galleries, and museums employed similar approaches to display and foregrounded consumption of art, fashion, and the cultural kudos attached to each. For example, Louis Vuitton sponsored a party for the launch of its spring/summer 2008 collection of handbags decorated with prints by Richard Prince. The party was held at the Guggenheim Museum in New York, on the last night of Prince's exhibition there, drawing comment from some areas of the press on the problems of commercial sponsorship and the status of fashion in the gallery. This demonstrated how art and fashion, although inextricably linked, can both gain and lose from comparisons made when they are brought into close proximity.

Miuccia Prada has been very active in examining these cross-currents. In 1993, she established the Fondazione Prada to support and promote art. She also commissioned architects, including Rem Koolhaas, to design iconic 'epicentre' stores for her, which would provide a space for art exhibitions to be held alongside her clothing on the shop floor. This included huge photographic prints by Andreas Gursky at her store in Soho, New York. The fact that Gursky's work has frequently critiqued consumer culture adds an ironic edge to Prada's display of his photographs. Thus, architect, artist, and designer are presented as knowing and self-aware, creating fashion, art, and buildings, while simultaneously commenting on these practices.

Miuccia Prada's complicated relationship with fashion and art was best expressed in her exhibition *Waist Down: Miuccia Prada, Art and Creativity*, which examined the evolution of skirt design

8. The 2006 touring exhibition *Waist Down* included creative displays of Prada skirts

within her collections. Designed by Koolhaas' architectural team, the show travelled internationally, held in venues such as the Peace Hotel in Shanghai in 2005. The exhibition used experimental display methods; skirts hung from the ceiling on special mechanized hangers which spun them round, or were spread out and encased in plastic to look like decorative jellyfish. Prada's financial acumen and global success enabled such innovative design to be possible, and her connections within the art and design world facilitated its realization.

However, Prada herself seemed intrigued by the ambiguity of these connections, and yet conflicted about how this linked fashion and art. When the exhibition travelled to her New York boutique in 2006, she commented to journalist Carl Swanson that 'shops are where art used to be', but went on to demur over the status of her exhibition and the other works displayed at her epicentre store, stating that:

> It's a place for experimentation. But it's not by chance that the exhibition is in the store. Because it started with the idea of putting more things to discuss, mainly about my work, in the store. It's like an explanation of the work. It's not at all anything connected with art. It's just to make the store more interesting.

This contradiction lies at the heart of fashion's relationship with art. Collaborations between artists and fashion designers can produce interesting results, but there can be discomfort from both sides about how such work is perceived. As important aspects of visual culture, fashion and art both represent and construct ideas about, for example, the body, beauty, and identity. Nevertheless, art's commercial side is revealed by its closeness to fashion, and fashion can seem to be using art to provide it with gravitas. What is revealed by such crossover projects is that each medium has the potential to be both consumerist and conceptual, meaningful and about surface display. It is these similarities that bring fashion and art together, and which add interesting tensions to their relationship.

Chapter 3
Industry

The 1954 British short film *Birth of a Dress*, directed by Dennis Shand, begins with a shot of London store windows filled with fashionable ready-to-wear dresses. As the camera pans across the shiny surface of the windows, a voiceover comments on the diversity of fashions available to British women, and the role of haute couture as inspiration for readymade designs. The frame then closes in on a specific cocktail dress; fitted close to the figure with a deep flounce down one side, it expresses the verve of 1950s eveningwear. The dress, we are told, was designed by noted London couturier Michael Sherard, and then adapted to become a mass-produced garment, available to 'the ordinary woman on the street'. The film then details this process. The fashion media usually work to cover up the industrial background from which clothing emerges, but *Birth of a Dress* positively celebrates the wonders of British manufacturing and design, which have gone into the dress's production. Sponsored by the Gas Council and Cepea Fabrics, it unmasks the series of factories where the cotton for the dress is bleached and prepared. The viewer is taken inside the textile mills' artists' offices, where the fabric print is designed, in this case a typically British floral of a rose sketched in charcoal. The etching process that transfers the print to a roller, the science laboratory that develops the aniline dyes (a by-product of the gas industry), and the factory printing mile after mile of the fabric are

9. A still from the 1954 film *Birth of a Dress*, which tracks the design and production process of a mass-market dress range

all proudly displayed as evidence of the North of England's expertise and invention.

The focus then shifts to Michael Sherard's refined Mayfair salon, where, inspired by the fabric, he fashions an original evening dress. From there, a Northampton factory's ready-to-wear designers reinterpret the dress for the mass-production process. Simplifying the design, they produce a stylish gown in three colour ways that is presented in a fashion show to international buyers. Thus, the viewer is reminded of the varied stages necessary in the production of the fashionable clothes she wears. The design is connected to British success in couture and mass fashion, and the viewer is prompted to see these clothes as 'allied to all that is newest in industrial research and scientific development'. The film is a post-war promotion of industry, Britishness, and burgeoning consumerism. Its focus on the process that goes into fashion's creation was unusual, since it connected all aspects of an industry which is normally presented only in fragments: as a complete garment, a designer's idea, or an object to aspire to.

As *Birth of a Dress* shows, the fashion business comprises a series of interconnecting industries. At one end of the spectrum these focus on manufacturing, and at the other on the promotion and dissemination of the latest trends. While producers contend with technology, labour, and managing the commerce of design, journalists, catwalk show producers, marketers, and stylists turn fashion into spectacle and make trends comprehensible to the consumer. Clothing is transformed by these industries, literally through the manufacturing process, and metaphorically through magazines and photographs. The fashion industry therefore produces not just garments, but also a rich visual and material culture that creates meaning, pleasure, and desire.

In their article on the industry's development, Andrew Godley, Anne Kershen, and Raphael Schapiro have shown that fashion is predicated on change. It is inherently unstable and seasonal, and each facet of the industry therefore searches for ways to temper this unpredictability. Forecasting trade shows project several years ahead to set trends in textiles, and themes that can guide and inspire fashion producers. Brands employ experienced designers, whose instinct for evolving trends is balanced with signature pieces to create successful collections. Fashion show producers and stylists then present collections in the most enticing way to develop the label's image, gain press coverage, and encourage stores to place orders. Store buyers rely on their awareness of their customer profile and retail image to purchase the outfits most likely to sell well, and reinforce the fashion credibility of the retailer they represent. Finally, the fashion media from style magazines to high-fashion titles advertise and editorialize fashion to seduce and entice their readers.

Fashion's development since the mid-14th century has been based upon technical and industrial breakthroughs, tempered by reliance on long-standing traditions of small-scale, labour-intensive methods that retain the flexibility necessary to meet the challenges of seasonal demands. Importantly, the fashion industry is also driven by consumer demand. During the 18th century, there was a shift from annual changes in textile designs and fashion styles to seasonal changes. Wearers would adapt their clothing in line with seasonal trends, to create new effects through trimmings and accessories. While the wealthy could afford expensive bespoke fashions, as Beverley Lemire has noted, those lower down the social scale could combine second-hand and, from the 17th century, readymade garments.

Clearly, fashion went beyond a process of simple emulation, either of aristocrats, or later of French couture styles. While it should not be assumed that everyone did, or for that matter could, follow fashion, consumer demand is a significant factor in its advance as an industry. Since the Renaissance, aspirations to individuality, aesthetic sensibilities, and the pleasure taken in clothing, whether tactile or visual, all played a part. The industry therefore generates local, national, and international fashions, with makers and promoters catering to diverse desires and needs. From the regional fashions of young, 18th-century English apprentices, eager to distinguish themselves through the trimmings on their clothes, or the elaborate velvets of 16th-century Florentine dignitaries, fashion involved a complex chain of traders, distributors, and promoters.

The evolution of the fashion industry

The Renaissance industry thrived on a global trade in fabrics, with free cross-pollination from East and West. Garments were made up using gradually more sophisticated methods, improved in the 16th century by Spanish tailoring books that enabled better fit. Wars and trade led to styles spreading across the Western world in the later 15th century, the Burgundian court's etiolated styles dominated, while dark Spanish fashions spread in the following century. Such fashions were part of consumers' desire for luxury and display, which was formalized in the 17th century by Louis XIV's regulation of the French textile trade. While this consolidated a centuries-old textile manufacturing and global trading network, the French monarch's efforts also recognized fashion's role in shaping not just a nation's identity, but also its economic wealth. This imperative later saw the formation in 1868 of what would become the Chambre Syndicale de la Haute Couture, to police the haute couture industry in Paris. At the other end of the scale, newly industrializing countries continue to build up their own fashion and clothing industries, as witnessed by Mexico's upgrading of its production capacity during the 1990s.

The 17th century saw growing recognition and consolidation of rich fabrics in Lyons, luxury trades in Paris, and tailoring in London, based on small-scale making-up of garments, frequently carried out in little workshops or households that focused on traditional craft skills. While this encouraged wealthy locals' and tourists' consumption of fashions, it was the early attempts to make readymade clothes that would lead to the fashion industry's wider impact, in terms of dressing more people, increasing financial gains, and, ultimately, in its status as a major international economic and cultural force.

Military needs drove significant advances in the readymade industry. The Thirty Years War (1618–48) saw the development of a large standing army, uniformed by both military and contracted-out workshops, a process that increased during the 18th century and later Napoleonic Wars. Early readymade garments focused on nondescript dress, clothing sailors in 'slops', the wide-legged breeches they commonly wore, and basic garments for slaves. While this was not part of the fashion industry *per se*, it set the necessary prerequisites for the ready-to-wear industry which was subsequently to emerge.

America's development as a nation played a crucial role. In 1812, the United States Army Clothing Establishment opened in Philadelphia, one of the earliest readymade manufacturers. Along with the huge demand for uniforms during the Civil War, and Levi Strauss' Gold Rush-driven denim business, an industry was emerging based upon greater standardization of methods and garment sizing. Claudia Kidwell identifies a parallel change in attitude towards readymade clothing in the later 19th century. It was no longer seen as denoting lack of money and status. As urbanization increased, city workers and dwellers wanted affordable clothing, 'which looked in no way appreciably different from the mainstream fashion'. The greater visibility of fashions in the city, and people's corresponding desire for individuality amidst the crowds, was another motivating force for the industry.

Demand was interconnected with innovations. The spinning jenny (c. 1764) speeded up textile production, and the jacquard loom (1801) increased the complexity of fabric designs. However, it was the development of a rational sizing system that allowed effective mass clothing production, and the growth of the broader fashion industry from the mid-19th century. By 1847, for example, Philippe Perrot states that there were 233 ready-to-wear manufacturers in Paris, employing 7,000 people, while in Britain, the 1851 census showed that the clothing trade was second only to domestic service as the largest employer of women. By this point, readymade womenswear was also developing and, as with early menswear examples, it focused on easy-fitting garments such as mantles.

Singer's introduction of the sewing machine in 1851 is sometimes credited with revolutionizing ready-to-wear. However, it was not until the 1879 invention of an oscillating shuttle running on steam or gas that a marked difference was made in the speed and ease of manufacturing. Andrew Godley has written that a skilled tailor could make 35 stitches per minute. However, by 1880 powered sewing machines could produce 2,000 stitches, and in 1900 this figure had increased to 4,000 stitches per minute. Further innovations, in cutting and pressing techniques, for example, reduced costs to manufacturer and consumer, as well as production times.

Immigrants fleeing the pogroms in Russia in the 1880s added further impetus to the British and American readymade industries, and Jewish tailors and entrepreneurs played a fundamental role in the fashion industry's development. Elias Moses, for example, staked his claim in advertising as 'the first House in London... that established the system of NEW CLOTHING READY-MADE', further asserting that 'tailoring is as rapid in these days as railway travelling'. Moses' association of his own trade's speeded-up methods with faster modes of travel is apposite. Not only did the train system quicken trade and distribution, it opened up the potential for travel, spreading fashions across and between classes as well as countries.

Travel and holiday clothes, sports and leisure fashions, from black veil 'uglies' to shield women from seaside sunshine in the mid-century to the steady rise of the more relaxed lounge suit for men, powered the growth of readymade fashion. In the last quarter of the 19th century, women's entry into white-collar work necessitated new styles appropriate to the public sphere. 'Tailormades', the prototype of women's suits, developed in the 1880s. Worn with blouses, they represented yet another option in the burgeoning array of fashions opening up to both sexes at the end of the 19th century. Indeed, the American 'shirtwaist' blouse became a huge craze in the early 1890s, and showed the close alliance between consumer demand and supplier innovation that motored the fashion industry.

If the 18th century had witnessed a growth in Western consumer culture that sparked people's desire for fashion, then the 19th century turned this love of novelty and sensuality into a frenzy of spectacle and commerce that spread across the globe. Inventors patented a quick succession of mass-produced crinolines, corsets, and bustles to reshape women's bodies using the latest technologies; rubber and celluloid provided collars and cuffs to young men eager to adopt the white-linen elegance of a gentleman cheaply and easily, and aniline dyes meant fabrics brazenly combined fashion with scientific innovation.

At the same time as this acceleration within the readymade industry, couture was adopting increasingly astute business methods. Promotional techniques, especially fashion shows, employed to great effect by, for example, Lucile and Worth, as well as leading department stores, disseminated elite visions of fashion style. These generated publicity at all levels of the market, and provided templates for manufacturers eager to adapt the latest trends to their own price point. American buyers were particularly keen to take advantage of the commercial potential of couture's aura of authenticity. They paid to attend shows, purchasing a pre-agreed number of garments, from which they could produce a limited number of copies. As in the 17th century, Paris was a synonym for luxury, the city's name exploited in advertising and editorial copy, and attached to shop and brand names internationally as a marker of fashion credibility. Paris embodied elegance and Old World luxury, and it also provided a model for other cities' clothing industries, as each sought to formulate its own saleable signature for the domestic and international market.

By the end of the 19th century, fashion's growth as a driving force within the clothing industry brought stylish clothes to a wider cross-section of people. While fashion enabled people to construct new identities, its under-side was the exploitation of workers, usually female and frequently immigrant. The sweatshop was a dark shadow haunting the industry's burgeoning modernity. From the 1860s onwards, reports shocked both governments and public with tales of the cramped conditions, long hours, and poor wages that kept retail prices down and enabled deadlines to be met. Debate over the ethics of production led to greater unionization and, in the early 20th century, laws concerning minimum wages. While it is the clothing industry, with its focus on mass-produced, standardized garments, which has been guiltiest in its exploitation of labour, fashion continues to cause controversy. The Victorian image of emaciated young women sewing couture gowns has been replaced by exposés of brands using child labour in Asia and South America.

While fashion manufacturers had traditionally needed to be close to the market, to respond quickly to consumer demand for particular trends, better information systems meant making up could be subcontracted to increasingly far-flung sites. As the 20th century wore on, technology enabled sales figures for each garment style to be collated from shops' individual cash registers to enable orders to be made rapidly. Improved travel and distribution speeded up this process further, aiding internationally successful brands such as Sweden's H&M, and Spain's Zara. Such companies could reproduce, and in some instances pre-empt, high-fashion trends by responding both to designer collections and close observation of emerging trends on the street. It also meant that it was harder to ensure working conditions, leading to accusations against high-street brands such as Gap.

By the 1930s, the structure for the contemporary fashion industry had already been established. As the century wore on, it would become known as 'Fast Fashion', as it came to supersede the industry's previous seasonal timetable with regular supplies of new garments sent out to high-street retailers. The boom years of the 1920s were pivotal to establishing the foundations of this system. The decade saw greater investment and international communication, as well as increasing evidence that fashion, rather than quality or function, could be used to sell products from clothing to cars. As the Depression set in, cutbacks led to a focus on streamlining industrial practices, building domestic markets, and seeking out new global regions to target, with Parisian couturiers and American ready-to-wear manufacturers both identifying South America as an important potential source of new customers.

The post-war period saw further consolidation of markets as well as domestic industries. American backing and business know-how aided the Italians and Japanese to develop their industries with a balance between fashion-led garments and wardrobe basics. Indeed, so important is this combination that Teri Agins identified it as crucial to a label's business survival. She asserted that American designer Isaac Mizrahi had to close his eponymous business in 1998, because he had focused entirely on fashion garments and ignored the need for classics.

Once again, this demonstrates the fashion industry's volatility, and designers' and manufacturers' need to factor in ways to increase and stabilize their market share. This can be seen in couturiers' establishment of licensing deals and ready-to-wear lines, and in the late 20th century, ready-to-wear label diffusion lines, such as Junior Gaultier and DKNY. These collections play upon the designer's aura, already established in their main lines, to widen their customer base with more affordable and usually more basic garments.

The need for outside investment and other means to financial assurance were tackled over the course of the 20th century. Burton's menswear manufactured and retailed its own designs, allowing a close relationship between demand and supply to be nurtured, and enabling the company to go public in 1929. From the late 1950s, French fashion labels were floated on the stock exchange. Since the 1980s, luxury giants, such as Louis Vuitton Möet Hennessey, whose portfolio includes Marc Jacobs, Louis Vuitton, Givenchy, Kenzo, and Emilio Pucci, ensured fashion credibility by grouping together younger labels with established houses, while protecting against losses by spreading profits across a wide range of wine, perfume, watches, and fashion brands.

However, there is still a significant segment of the fashion industry that continues to work on the same small-scale, labour-intensive model which has survived for centuries. This is epitomized by the studio-workshops mainly focused in London's East End, where young designers such as Gareth Pugh, Christopher Kane, and Marios Schwab employ a tiny number of assistants to enable them to produce their collections. In a tradition set by British designers since the 1960s, the strong fashion content of their work attracts press interest and spreads their influence globally.

The development of fashion promotion and dissemination

The fashion media and promotions industry has developed in tandem with manufacturing and design, disseminating information on new fashions, and constructing ideals of fashion through imagery and text. While press coverage can undoubtedly boost designers such as Pugh, Kane, and Schwab, it can also undermine longer-term development. If young designers gain too much notoriety very early in their careers, before they have gained sufficient financial backing and manufacturing capability to fulfil orders, it can be hard for them to develop their businesses. However, press coverage is viewed as crucial to building a profile and, ultimately, to finding economic investment from a reliable backer. This contradictory situation has particularly plagued London Fashion Week, where art schools such as Central Saint Martins School of Art and Design regularly produce talented designers, but lack of infrastructure and government investment leaves them vulnerable.

In the second half of the 20th century, a cycle of seasonal international fashion shows came to dominate the industry. These provided a platform for designers and manufacturers to display their collections as they wished them to be seen, rather than through the filter of magazine coverage. Fashion shows brought together buyers, whether from international stores or, in the case of couture, wealthy individual clients, and, as they developed, members of the press and photographers. From tiny showings in couture salons in the late 19th century, the catwalk show evolved its own visual language, comprising the models' movements and gestures, lighting and music accompaniments, and increasingly elaborate performances designed to convey each label's signature and vision.

Until the 1990s what was seen in these shows was filtered to the public through other media, whether newspapers, magazines, or later television channels such as Fashion TV. However, in the late 20th century, the Internet provided the general public with access to unedited shows, sometimes broadcast simultaneously on the designer's website. This immediacy has the potential to alter the balance of power between designers and manufacturers, the fashion media, retailers, and potential consumers. It brings an unmediated version of the designers' work to customers, who can demand items seen on the catwalks, which have not necessarily been picked up in magazines or by store buyers.

The international network of print, broadcast, and online media, reliant upon dramatic imagery to create fashion meaning, has evolved over centuries. During the Renaissance, trade and travellers, whether from local towns or abroad, would bring news of fashions. Caricatures mocked and celebrated fashions in equal measure. Leading dressmakers contrived to spread trends by sending out dolls dressed in the latest formal and informal styles. Letters provided an informal means to communicate information on new styles. Indeed, Jane Austen's correspondence with her sister Cassandra contains more fashion information than her novels do, detailing new trimmings on hats and new dresses purchased. This more anecdotal spread of fashions continues in online blogs and is mirrored in the intimate style of smaller magazines, such as *Cheap Date* which focuses on vintage and DIY fashions.

By the 17th century, more formal methods evolved, including irregular fashion magazines, which took their cue from earlier costume books that showed the clothing of different countries. However, it was not until the 1770s that the first regular fashion magazine appeared. *The Lady's Magazine* set in train a whole industry of fashion journalism and image-making. What is perhaps most striking is how the format of such magazines has remained a template into the 21st century. Fashion magazines of the 18th and 19th centuries combined gossipy social events coverage, which detailed society figures' outfits, advice on beauty and style, fiction, and news from Paris's leading couturiers and dressmakers. Fashion magazines contained a powerful combination of didactic articles and sisterly advice on appropriate fashions, beauty, and behaviour. They constructed ideals of femininity, whether strongly moralizing visions of dutiful domesticity, as seen in the *Englishwoman's Domestic Magazine* with its inclusion of advice on housework and paper patterns for dressmaking, in the mid-19th century, or avant-garde challenges to sexual identity as seen in *The Face* in the mid-1980s.

From early on, magazines had close ties with fashion houses and manufacturers, through advertising and, more insidiously, through promotional links. In the 1870s, *Myra's Journal of Dress and Fashion*, for example, pioneered the advertorial, bringing together advertising and editorial content, featuring articles written by Madame Marie Goubaud, as well as advertising, imagery, and editorial coverage of her fashion house. This relationship grew in the 20th century. In the 1930s, Eleanor Lambert was one of the first to apply public relations techniques to fashion, recognizing the possibilities of multiple types of promotion. Thus, press representatives lobby to have labels included in editorial text and imagery, adding the fashion kudos of the magazine to validate existing coverage in advertising. Lambert also encouraged film stars she represented to wear items by her stable of designers. In the 1950s, she sent sportswear designer Claire McCardell's new sunglasses range to Joan Crawford, in the knowledge that a photograph of Crawford wearing McCardell's designs would endorse the designer's work, while raising the star's fashion status.

Such cross-fertilization underpins the fashion industry. In the late 19th century, London's leading couturiers, such as Lucile, provided gowns for leading actresses to wear on stage, gaining free publicity and increasing the visibility of their wares. This practice continued, with designers creating costumes for films, whether in the form of iconic couture gowns by Givenchy for Audrey Hepburn in *Breakfast at Tiffany's* in 1961, or the cartoonish, sci-fi excess of Jean-Paul Gaultier's costumes for *The Fifth Element* in 1997. Crucially, actors and celebrities wore fashions during their 'private' life that helped to promote the idea that a particular designer's work connected intimately to their lives. Global coverage of events such as the Academy Awards ceremony provoked designers to compete to lend stars their gowns. While magazines, including *Hello*, followed in the footsteps of earlier Hollywood fan magazines to blur distinctions further between public and private by setting up shoots that show celebrities at home, listing the sources of everything they wear.

This interdependence between different strands of fashion and allied industries has been criticized as creating uniform ideas of acceptable identities. While this may be true to an extent, since the dominant image is undoubtedly slim, white, and youthful, fashion has simultaneously tested boundaries. Fashion and style magazines are part of the culture that spawns them, and therefore reflect wider attitudes towards race, class, and gender. Their role in representing what is new, and the fact that they can attract leading writers and image-makers, also means that they can suggest new identities and create a pleasurable escape from the everyday. In the 1930s, American *Vogue* promoted the idea of the dynamic, modern woman, mixing more practical advice on what to wear to work with dramatic photo-shoots of aviatrixes, which suggested freedom and excitement. In the 1960s, British publication *Man About Town* brought together lifestyle advice for its male readers with imagery of smart suiting, photographed against stark urban exteriors. In Russian *Vogue* in the late 1990s, couture luxury and excess became a dreamscape from which to forget economic crisis.

Each publication formed its own style, to entice its audience and provide them with a marker of their own fashionable status. In the early 20th century, *The Queen* represented elegant, elite style; 1930s' *Harper's Bazaar*, under the editorship of Carmel Snow and art directed by Alexey Brodovich, created a magazine of high fashion, and dramatically paced pages of modernist elegance through its combination of strong text, imagery, and graphics; while *A Magazine*, produced in Antwerp since the 1990s, brought in avant-garde designers such as Martin Margiela to 'curate' each issue. Trade publications provide the analogue to the fantasy of much newspaper and magazine coverage, but are equally important in connecting fashion's disparate elements. The 19th-century publication *The Tailor and Cutter: A Trade Journal and Index of Fashion* provided practical information and technical discussion. Since the 1990s, websites, most significantly WGSN.com, have pooled information from global offices on trends predicted by international consultancies with coverage of what is happening on the streets of cities across the world, to enable the fashion industry to have instant access to emerging trends and developments.

The collages of image and text, body and clothing, editorial and advertising that fashion magazines created produced a space that readers could escape into. They constructed a realm of visual consumption, where even the feel of their pages, whether the glossy sheen of *Elle* or the textured inserts of *Another Magazine*, contribute to a multi-sensory experience. Although they are ephemeral, they are documents of contemporary culture and society, and unite the commercial imperatives of the fashion industry with its intangible role in global visual culture. Not only do they report fashions, for many people they *are* fashion. The meanings that illustration and photography add to garments in some cases transform them into fashion. Between the everyday reality of clothing and the vision created through an illustrator's interpretation or the alchemy of a fashion shoot, layers of new ideas are brought to bear. These tap into contemporary mores, but frequently go far beyond what already exists to suggest heightened reality or surrealist narratives.

In this early 19th-century fashion plate, the illustrator simplified the lines of his sketch, echoing the purity of the fashionable silhouette. By showing the main figures in back view, focus is given

10. An 1802 fashion plate that shows the classicized fashions of the period

to the antique referenced drapery of the woman's dress, which is emphasized by the sweep of her rich red shawl. The shrunken tails of the man's coat are also stressed, set against the classically inspired 'nudity' of his flesh-toned pantaloons. Other fashion details, from the men's modish sideburns, to the seated woman's little scarlet hat, are set within the illustration's narrative. Fashion plates added mood and context to clothes, enhancing the raw information of simple illustrations that acted more as a template to show a dressmaker or tailor when ordering an outfit. The environment created a feeling of relaxed elegance, and connects clothing to wider fashions, in this case contemporary fascination with hot air balloons.

Fashion photography, which developed from the mid-19th century, performed a similar function, with the added element of showing clothing on real bodies. If production seeks to counterbalance fashion's unpredictable nature, then fashion imagery celebrates its ambiguities. Representation has played a central role in fashion's formulation, showing how styles might look on the body, and cataloguing the movements and gestures associated with particular garments.

This 1947 image by American photographer Toni Frissell shows how simple, everyday clothes can be transformed through representation. Rather than showing this tennis outfit in its usual courtside setting, Frissell places the model against a dramatic mountainous landscape. Natural lighting makes its bright white fabric glow, the crisp silhouette sharpened by sunshine. The model remains an anonymous identifying figure for the viewer. She turns away to look at the view, her pose emphasizing her athletic figure, but not far removed from a natural gesture. The balcony's curve connects her to streamlined modern architecture, and situates her in an environment speaking of both natural and manmade luxury. The fashion editor's choice of model, and styling of the shoot, with clean plimsolls and ankle socks, simple hair grip, and casually discarded cardigan, add to the idea of nonchalant ease projected by

11. Toni Frissell's 1947 fashion photograph of a model in tennis dress set against a dramatic landscape

Frissell's staging and composition. Thus, ready-to-wear garments are given a gloss of fashionable grandeur they might otherwise lack.

The interconnecting industries that intercede between makers and consumers therefore create various points at which 'fashion' appears. These are incremental and cumulative. John Galliano's fashion training, experience, and intuition mean that his initial sketches contain future fashions, which are then amplified through the process of their evolution. The skilled craftspeople he works with in the Dior ateliers further contribute to a centuries-old tradition of couture fashion credibility. At his catwalk shows, his fashion statement is brought to industry insiders through elaborately dressed environments and theatrical deployment of models and styling. The fashion press then reinforces and, potentially, reinterprets Galliano's fashion vision through written descriptions of key trends, connecting his work to that of his peers. Advertising and editorial photographs, retail and window displays, all act to validate his work as fashion, and suggest ways to imagine how it might be worn.

It is hard to single out the point at which clothing becomes fashion. In the case of couturiers such as Galliano, or earlier examples such as Balenciaga in the mid-20th century, it was through their working practice, but also via the constellation of promotions and advertisements through which their designs were mediated. For ready-to-wear and high-street stores' lines since the 1930s, it has been a similar mix of established fashion credibility built up over time, validation by the media, and an intangible ability to express diverse inspirations through dress in a way that connects clothing and body ideals to other aspects of contemporary culture.

Chapter 4
Shopping

In 2007, Comme des Garçons opened a new 'guerrilla store' in Warsaw. Scheduled to remain there for just one year, it was part of a programme of similar 'pop-up' shops by the label; the first was in East Berlin in 2004, followed by similarly transitory boutiques in Barcelona and Singapore. Each had its own character, in keeping with its environment. In Warsaw, the shell of an old Soviet-era fruit and vegetable shop remained intact, with green tiling, patchy plasterwork, and traces of ripped-out fittings on the rough walls. This aesthetic was extended into the 'display cabinets', really Soviet furniture, installed to house the label's range. Cabinets clung haphazardly to the walls; drawers spilled out, lopsided and half open to expose shiny perfume bottles; broken chairs cascaded from the ceiling with shoes balanced precariously on their battered seats; clothes were hung on bare metal rails; and twists of wire hung from light fittings and curled on the floor, half hidden under the stacks of fittings.

The effect was of an abandoned storeroom, with clothes and accessories left behind in the shopkeeper's rush to leave. This atmosphere was symbolic of its geographical and historical context, with communism abandoned in former Soviet bloc countries, to be replaced by capitalism. This has led to a shift from buying what was needed, or rather what was available, to shopping

12. **The interior of Comme des Garçons' 2008 guerrilla store in Warsaw is designed to look like a modernist furniture exhibition**

for what is desired and aspired to, from a wide choice of goods. The rawness of the shop also chimed with the nature of guerrilla stores, which suddenly take over an urban space. Indeed, this was the label's third incarnation in Warsaw, the first had appeared in 2005 in a derelict passageway under a bridge.

Although they might seem unplanned, such shops are part of Comme des Garçons' strategy to remain at the forefront of fashion retailing. Some of the stores remain open for only a few days, others a year; none are advertised, other than through emails to existing customers, perhaps a few posters in the local area, and, crucially, through word of mouth. These processes mimic the effects of a subculture, reaching out to opinion-makers within an inner circle already aware of the label's status in the fashion industry as pioneers of avant-garde style and design. The guerrilla store creates an atmosphere of exclusivity, intrigue, and excitement around its products. It promotes the feeling that its visitors have privileged knowledge, and that they are taking part in a semi-covert event by shopping there. It therefore plays into the key elements of early 21st-century high-fashion consumerism, by emphasizing desire, lifestyle, and identity. As such, the store, again like street cultures, suggests individuality yet membership of a group. It advocates shopping as an experience, in this case akin to visiting a small art gallery. Importantly, it builds the brand in a manner that is in keeping with its intellectual ethos. It apparently rejects the excesses and decadence of much fashion advertising and retailing, while remaining a shrewd marketing device to target its core audience, as well as luring in the curious passer-by.

Since the 1980s, Rei Kawakubo, the designer behind Comme des Garçons, has launched a series of innovative shops. The spare, minimal spaces of her early boutiques drew upon the aesthetics of traditional kimono shops, with garments folded on shelves. This was combined with a reverential air produced by the limited number of items on display, making shoppers focus on details and packaging. Her peers, as well as high-street brands such as Gap and Benetton, mimicked this approach, with wooden floors, plain white walls, stacks of sweaters piled on shelves, and carefully positioned clothes rails that emphasized space and clean lines.

Dover Street Market in London opened in 2004 by Kawakubo and her husband Adrian Joffe took a different approach, with carefully presented fashion and design labels shown in separate spaces across the building. On one floor, a changing room is housed in an oversized gilded birdcage, on another clothes are grouped with plants and garden accessories. Kawakubo's conception of Dover Street Market is as a place that is flexible and varied; she states on its website that:

> I want to create a kind of market where various creators from various fields gather together and encounter each other in an ongoing atmosphere of beautiful chaos: the mixing up and

coming together of different kindred souls who all share a strong personal vision.

The affect is of a contemporary version of a 19th-century bazaar, populated by a changing array of exclusive fashion lines and eclectic objects.

Alongside Comme des Garçons' more permanent boutiques, such enterprises stress the importance of variety and flexibility in modern retailing. In a saturated market, designers and fashion labels of all kinds must distinguish their identity to build a strong customer base. While Comme des Garçons represents the cutting edge of this enterprise, its methods hark back to earlier predecessors, from 19th-century department store entrepreneurs who understood the need to create spectacle around their goods, to early 20th-century couturiers, who designed their salons as intimate sensual spaces that mirrored the style of their clothes.

The development of retailing

During the Renaissance, fabrics and trimmings were, as they had been for centuries, bought from markets and a range of itinerant pedlars. Lace, ribbons, and other decorative items would be taken around the countryside, or sold wholesale to local stores. Larger villages might have a draper's shop, which sold wools and other materials, while towns might also have a milliner's, which would sell the finest silks and wools. Local dressmakers and cobblers would make up clothes and accessories, and buying garments could therefore be a lengthy process, as the elements of an outfit were bought from various shops and then made up by craftspeople. Purchasing patterns were different in each country. In England, people would often travel to a nearby town or city to buy more fashionable clothes. However, the fragmented politics and geography of Italy meant greater distinction between regions, and therefore a wider range of shops in each village.

A global trade in textiles had been established for millennia, with international routes crossing Asia and the Middle East into Europe. Huge fairs were held to buy and sell fabrics to merchants and pedlars who would travel to, for example, Bruges or Geneva, or later Leipzig, where fairs were held three times a year, or to Brigg market in Leeds. During the 17th century, the English and Dutch East India Companies (EIC) improved trade links with Asia. By the mid-18th century, cotton from India, for example, became an everyday fabric. It was fashionable and, more significantly, it was cheap and washable, and therefore brought greater levels of cleanliness to people of all classes. Such goods could be transported across the globe because of improved shipping. There was also an increasing demand for fashionable textiles, as more people wanted to be stylish and respectable, conforming to contemporary ideals of appearance and behaviour. The EIC fed people's desire for new and changing textile designs, importing silks, cottons, and calicoes. Merchants spread new fashions by encouraging fashion leaders to wear their latest goods to stylish social events, which would then be reported in fashion magazines. Woodruff D. Smith has described how the EIC then commissioned Indian craftspeople to create more of the most successful designs, selling them on across Europe as the fashion spread out from Paris. As Daniel Roche has noted in relation to changes in dress in France, by the end of the 18th century, there was in general a far wider range of consumer goods available, 'but everything that related to the expression of appearances, both social and private, increases still more'.

Textiles and clothing were relatively expensive, given to household servants as part of their wages, passed down through families, and sold on through a chain of used clothing shops and markets until they fell into rags or were turned into paper. In the 18th century, with better agricultural practices and distribution of wealth, more people wanted to buy fashionable clothes, at the very least for Sunday best. Shopkeepers began to take more time over the display and presentation of their wares and in their approach to customers. By the 1780s, plate glass windows led to enticing

13. London street markets have traded in second-hand clothes for centuries

displays and interior displays were beginning to be more sophisticated. Fashionable shopping was already shaping the geography of cities. In London, Covent Garden had become the first fashionable suburb, with Inigo Jones' piazza housing various drapers' and milliners', which had moved west after the Great Fire of 1666. In Paris, the Palais Royal had been remodelled to provide perhaps the first purpose-built shopping centre, with rows of little shops and cafés around the perimeter of its gardens. Advertising and marketing were also developing. Handbills boasted of a particular shop's range of readymade garments or rich selection of fabrics; fashion magazines gave detailed descriptions and illustrations of the latest modes, and entrepreneurial manufacturers and salesmen encouraged fashion leaders to be seen wearing their goods. Since the Renaissance, shopping had developed hand in hand with a growing sense of personal identity. Fashionable dress provided the means to express this visually, and knowing where and how to shop for fashion was key to achieving this. Novelists such as Tobias Smollet satirized people's attempts to dress attractively, fashionably, and, frequently, above their station, conscious of the growing consumer culture that was to flourish in the following century.

The growth of shopping

In the early 1800s, small specialist shops continued to be important, but it was the emergence of larger establishments that began to group together a wider range of goods and services, which was to herald a new era in shopping. Aristide Boucicaut opened his Bon Marché in Paris in 1838, which by 1852 had evolved into a department store. It brought together fabrics, haberdashery, and other fashionable products, and introduced a strong social element to shopping by including a restaurant. Boucicaut developed various customer services, which added to the sense of a change in relationship between shop workers and customers, and between customers and the way they used a shop. His prices were fixed, and marked on all goods, which eliminated the need to haggle, and he also allowed refunds and exchanges. The Bon Marché was one of a number of early department stores, including Kendel Milne in Manchester, which had evolved from a bazaar in 1831, and A. T. Stewart in New York, which gradually changed from a small draper's in 1823 to hold the dominant position in the city's main fashion shopping area on Broadway by 1863. These stores evolved increasingly sophisticated sales techniques. Shoppers were encouraged to browse, following the carefully designed routes through the shop floors, visiting the cafés and restaurants there, or stopping to watch the entertainments that were provided. For the first time, shopping became a leisurely pursuit, focused upon spending time and, it was hoped, money, in a fashionable, secure environment.

Women were the main targets for the department stores, and were enticed into these elaborate buildings by carefully organized window displays which emphasized the play of light on fine fabrics and the rich colours and textures of their stock. Previously, it had been impossible for middle- and upper-class women to go shopping alone. Even with an accompanying maid or footman, certain streets were out of bounds at particular times of the day. Bond Street in London, for example, was a focus for shops for gentlemen, and it was improper for ladies to go there during the afternoons. These careful rules of etiquette were eroded by department stores, which encouraged women to socialize and browse, in what Edward Filene, the owner of a store in Boston, is quoted by Susan Porter Benson as calling an 'Adamless Eden'. Not only did this give women greater freedom, it also shaped them as consumers. Erika Rappaport describes this change in ambiguous terms. Victorian women were expected to be concerned primarily with family and home. Female shoppers could be seen as focusing on such domestic matters by buying items for their children and husbands, as well as fashionable dress for themselves, which would demonstrate the status and taste of their families. However, going shopping also meant leaving the privacy of the home, and visiting urban centres, the public sphere previously dominated by men. Shopping also focused on sensual experience, rather than more virtuous feminine occupations. In Rappaport's words, this was part of the development of the city as a 'pleasure zone', in which 'the shopper was designated as a pleasure seeker, defined by her longing for goods, sights, and public life'. Fashion therefore offered a contradictory experience. Shopping for clothes, accessories, and haberdashery allowed women to occupy a new space in the growing urban landscape of the 19th century, but it also potentially led them into a lifestyle focused on adornment and desire. Store owners worked to make their displays as seductive as possible, to persuade women to indulge themselves and spend whole days within their walls, or moving between the various shops that clustered close by in all large towns and cities.

Each store developed its own character, aiming to draw in customers who were attracted to their style, as well as to the diversity of their goods. Thus, in 1875, Liberty opened in London, selling furniture and objects from the East, alongside 'Aesthetic' dress, historically inspired loose gowns that offered an alternative to tightly corseted mainstream fashions. Some stores opened branches in other towns or in the suburbs, including, in 1877, Britain's first purpose-built department store, the Bon Marché in Brixton, South London. Other stores launched branches in stylish seaside resorts, including Marshall and Snelgrove's Scarborough store, which was open during the holiday season. The spread of department stores brought fashionable goods to a wider range of people. Most department stores had their own dressmaking departments, as well as selling the growing array of readymade clothes becoming available in the second half of the 19th century. Stores worked hard to build up a relationship with their customers, winning their loyalty through services, quality, and price.

These developments not only changed the ways in which people could buy fabrics and clothing; it simultaneously shaped ideas about how to behave and how to dress. Store advertising suggested acceptable standards of taste, and promoted an ideal of fashionable identity. This built on the increasing dissemination of fashions and desire to be part of consumer society, which was already established at the start of the century. Although department stores embodied bourgeois ideals, they embraced a wider range of people. In 1912, Selfridges, established in London along American lines and branded with its own shade of green carpeting, stationery, and delivery vans, opened a hugely popular 'bargain basement'. The open design of department stores allowed a wide range of people to come in and look around freely. Although grander shops may have intimidated some shoppers, others would save up for a luxury item from a store whose clientele's status and style they aspired to. By the 1850s, the growth of public transport made shopping trips by bus or train simple and affordable. Underground trains in major cities would make this process even easier and encouraged the idea of a day's shopping as a pleasurable and easy source of relaxation and entertainment.

Stores worked hard to tempt shoppers with a combination of spectacular fashion shows that brought the glamour of French fashions to a wide audience and exciting new technology. In 1898, Harrods in London attracted a large crowd and much press coverage for introducing the first escalators to take people from floor to floor. While in the early years of the 20th century, American stores staged a series of Paris fashion shows, with real models parading through intricate stage sets, shimmering under specially designed electric lighting. The names of these extravaganzas evoke their atmosphere of decadence and excess. In 1908, Wanamaker's in Philadelphia held a Napoleonic themed 'Fête de Paris', complete with *tableaux vivants* of the French court. Meanwhile, in 1911, New York's Gimbels' had a 'Monte Carlo' event. Mediterranean gardens were built in the store's theatre, along with roulette tables and other props, to give an authentic feel of Riviera luxury to the thousands of people who visited.

While department stores brought fashion to the masses, opening in stylish shopping areas from Prague to Stockholm and Chicago to Newcastle, they were far from being the only source of fashion. The elite continued to frequent the court dressmakers and bespoke tailors they had gone to for generations. Tiny specialist emporia still thrived, often springing up in line with new fashions. For example, the early 20th-century craze for huge hats covered in feathers led to shops opening to sell ostrich plumes and other trimmings. Changing styles and faddish accessories also tempted male shoppers. In addition to luxurious shops selling jewellery and accessories to wealthy gentlemen were those targeting younger men, eager to spend money earned from the rash of new white-collar jobs. As with women's fashion, styles were spread by popular figures of stage and, increasingly, screen, as well as sporting heroes. A changing array of colours and patterns in ties and cravats, collar studs and cuff links would enliven men's suits each season.

Mail-order shopping was another important innovation, particularly in countries such as America, Australia, and Argentina, where the distances between cities made visiting shops in person more difficult. Department stores had their own postal sales sections, which capitalized on improving parcel mail and the introduction of telephones. Marshall Ward, based in Chicago, had perhaps the most famous mail-order service, its catalogues tempting Americans with the increasingly wide array of ready-to-wear fashions for the whole family. Improving transport methods also helped this trade, taking goods by carriers' carts and stagecoach, and then by rail.

By the first decades of the 20th century, therefore, consumerism had evolved to embrace a wide range of people of different sexes, ages, and classes. As mass-production methods improved during the 1920s, the selection of fashions and accessories available grew still further, and shops had to work harder to sell them effectively, in the face of growing competition. The already successful department stores and specialist shops were joined by 'multiples', an early form of chain store, which spread across Western countries. In America, branches of shops selling inexpensive fashions inspired by Hollywood stars' costumes gained national popularity. In the United Kingdom, Hepworth & Son, which had opened as a tailor in 1864, expanded to have menswear shops throughout the country, and is still trading, having evolved into Next, a chain store for men, women, and children. Multiple-branch shops had the advantage of central buying and administrative systems, which could keep prices affordable and manage marketing and advertising campaigns. They worked to produce a unified identity for their store designs, windows, and staff uniforms. While the dominance of chain stores by the second half of the 20th century led to accusations of homogeneity and, ironically, a lack of real choice for consumers, familiar brands reassured many customers by supplying the same type and quality of stock in each branch.

In contrast, couturiers continued to sell their designs in ways that combined centuries-old traditions with contemporary innovations. While clients were served and fitted individually, couture salons incorporated boutiques selling early incarnations of readymade lines, plus perfumes and luxury goods designed to please their elite customers. Both couturiers' salons, which were open only to private customers, and, during the show season, select store buyers and their boutiques used modern design and display techniques to demonstrate their fashion currency. In 1923, Madeleine Vionnet had her fashion house remodelled along sleek modernist lines, with classically inspired frescoes. While from the mid-1930s, Elsa Schiaparelli embarked on a series of Surrealist window displays which promoted the wit and fantasy of her designs. In each case, these artistic references related to the philosophy of their clothes and were echoed in their labelling, packaging, and advertising, producing a coherent house style for customers to identify with.

Couturiers needed to project an image of exclusivity which gave a luxurious aura to everything that bore their name. Although fashion was increasingly used as a tool to sell readymade clothing all over the world, many stores still felt that Paris was the key source of new styles. For example, American department stores and fashion houses sent buyers to the French capital each season to purchase a selection of 'models', outfits which they would be licensed to reproduce in limited numbers for their stores. These designs would have the highest fashion status in stores' collections, and would be supplemented by designs based more loosely on Paris-led trends, as well as an increasing number of styles by native designers that diverged from French diktats. Buyers thus played a crucial role, as they needed to understand the fashion profile of the stores they represented and the desires of their customers. It was crucial to keep an ever-changing array of fashions on the shop floor. In 1938, Kenneth Collins, vice-president of Macy's, addressed the Fashion Group, an organization dedicated to promoting fashion in America, stating that:

... it is one of the truisms of retailing that the difference between success and failure in the fashion business is dependent upon the ability of merchants rapidly to get into new fashions and just as rapidly to get out of them when they are on the wane.

This turnover of novel styles was fundamental to the fashion industry.

Big department stores such as Saks Fifth Avenue would have a number of lines targeted at different consumers. From 1930, it had its own luxurious creations designed by the owner's wife, Sophie Gimbel, under the Salone Moderne label, plus fashions she had chosen for the store in Paris. It then had various ready-to-wear lines, including sportswear and clothes aimed at young college girls, as well as comparable menswear styles. In combination, these collections built Saks' fashion reputation, demonstrating the store's taste and discernment in dressing the full scope of its customer base. These were sold in specially defined areas of the store to reflect their audience and purpose, and advertised in fashion magazines and newspapers at key points in the year to optimize sales.

During the Depression, many stores had to stop visits to Paris and became increasingly reliant on homegrown fashions. Despite the economic downturn, fashion magazines such as *Vogue* and *Harper's Bazaar* continued to carry advertisements for shops of all sizes. Columns such as 'Shop-Hound' in *Vogue*'s various national editions encouraged women to make shopping trips, mapping out the 'best' areas to visit and the chicest boutiques and department stores to go to. Designers and stores cultivated close relationships with the fashion media through their press representatives, who worked to obtain advertising and editorial coverage in magazines. These connections continued in subsequent decades. However, the Second World War, and the continuing deprivations it caused, interrupted the flow and availability of goods. Despite shortages and rationing in most countries involved in the conflict, the dream of consumer goods was held out in many countries as a morale-boosting vision of the future.

As economies recovered during the 1950s, new initiatives began to develop. One of the key examples of this was the growth of designer-owner boutiques that appeared in London by the end of the decade. These demonstrated how fashion could evolve from small-scale entrepreneurs who understood their audience and the kind of clothes they wanted to wear. In 1955, for example, Mary Quant was prompted to open Bazaar on London's King's Road by her own frustration with the contemporary fashion scene:

> I had always wanted the young to have fashion of their own... absolutely twentieth-century fashion... but I knew nothing about the fashion business. I didn't think of myself as a designer. I just knew that I wanted to concentrate on finding the right clothes for the young to wear and the right accessories to go with them.

Quant produced fun clothes: baby-doll dresses, corduroy knickerbockers, and fruit-coloured pinafores, which helped to shape the style of the period. She and her peers spawned imitators across the globe, eager to capitalize on the trend for youth-driven, mass-produced clothes. Quant also provided a template for future designer-retailers, who would develop global reputations by dressing emerging youth cultures. Vivienne Westwood and Malcolm McClaren's shop, also on the King's Road, changed its exterior and interior design, as well as the look of the clothes it sold, in line with evolving street styles. From Teddy boy-inspired suiting as Let It Rock in the early 1970s, through hardcore Punk aesthetics in mid-1970s Seditionaries and Sex, to its final incarnation as World's End, an Alice in Wonderland-style boutique with wildly sloping floor and backwards-running clock. Westwood's design and retailing style were part of the fluidity of subculture. Styles emerged and shifted as the music, street, and art scene they were connected with moved on. This flexibility created an exciting sense of community and currency around her store,

14. Zara's stores look like boutiques, and lure customers in with open frontages and carefully coordinated displays

promoted by the DIY ethos of subcultures. As with Quant in the 1960s, it demonstrated how like-minded shops could group together to generate business and consolidate the fashion reputation of an area. In the early 21st century, Alphabet City in New York saw a similar constellation of designer-makers opening up in close proximity.

Indeed, Spanish chain Zara, which is known for its mix of classic and catwalk-inspired pieces, has based its success on a strategic version of this more organic development of shopping areas. Since opening its first store in 1975, Zara has expanded internationally, overtaking its main rivals on the high street. Each store is designed to look like a boutique, with themed garments grouped together with accessories, suggesting possible outfits to consumers. The chain is owned by Inditex, which includes Massimo Dutti, Bershka, and Zara Home in its portfolio. Its strategy is to open a large Zara shop first, which acts like a flagship, visually stating the ethos of the label, then branches of its other brands are launched close by. This encourages shoppers to walk between the shops, buying from Inditex's different labels and seeing how the clothes, accessories, and soft furnishings sold in each complement one another. Allied to this is Zara's quick response to fashion trends, with a small design team and close-knit manufacturing system, which allows new styles to be spotted and rapidly translated into new garments that reach the stores soon after they have been identified.

Other international brands have relied on their own design teams' ability to create affordable fashions, combined with celebrity and high-fashion collections. H&M has commissioned a series of lines from designers including Viktor and Rolf, Stella McCartney, and Karl Lagerfeld, as well as music stars Madonna and Kylie Minogue. These collaborations usually last for a limited period only, creating huge media coverage, and swarms of shoppers queuing to buy each collection as it is launched. The success of this approach is similar to couturiers' ready-to-wear lines and licences in the 20th century. The aura of high fashion is used to enhance the status of various mass-market stores, from America's Target chain to Britain's New Look. Perhaps the most famous collaboration of this kind has been between model Kate Moss and Topshop, the British chain store that has led the way in high-street fashion since the late 1990s. This has seen an interesting exploitation of a star's personality, style, and aura of exclusivity into a regular range for the brand's branches across the world. These clothes mimicked items from Moss's own wardrobe of vintage and designer fashions. Moss herself is also a brand, used to market the range, and even to inspire decorative devices, including the twin swallow tattoos she has on her back which have decorated everything from jeans to blouses. This takes the connection between celebrity and fashion, which had been apparent since at least the 18th century, further than ever before.

This collaboration is demonstrative of the ever more blurred line between luxury and mass fashion since the late 20th century. In Britain, Kate Moss's collection is sold in Topshop's own high-street stores, and is therefore seen as part of a fashion-led, but undeniably mass-produced, world of throwaway fashions. However, in New York, the range was launched in exclusive fashion speciality store Barney's, giving it the air of an exclusive, luxury label that was sold alongside established high-fashion designers from across the globe.

This confusion between high and mass fashion is the result of the growing strength of ready-to-wear fashions over the past 150 years, and the strong fashion-led design ethos of high-street lines. As consumers have become more comfortable mixing vintage, designer, and cheap high-street and market finds together, the divisions between these categories has, to a certain extent, collapsed. Although prices still provide the most obvious difference, more emphasis is placed upon consumers' ability to put together an interesting and individual outfit than to adhere to fixed ideas of what is respectable. This change has not just come from the high street. Since the 1980s, luxury brands have extended their reach, moving from the elite confines of small boutiques to build huge flagship stores in major cities, as well as allowing their goods to be sold in duty-free shops and shopping centres specializing in knockdown price, old-season fashions.

In the late 20th century, luxury brands such as Gucci developed into huge conglomerates and quickly identified the Far East as the key market for their goods. Stores were opened both in mainland Japan and Korea, for example, but also in places where fashion-conscious people holidayed. Hotels in destinations including Hawaii hosted luxury boutiques where young, affluent Japanese women would shop. Blanket advertising campaigns balanced references to the exclusive heritage of brands such as Burberry and Louis Vuitton with a cutting-edge fashion image bolstered by the appointment of young designers, in these cases Christopher Bailey and Marc Jacobs. Online fashion stores, including high-end website net-a-porter.com, have made it even easier to purchase these fashions. Many of these follow a magazine format, with exclusive offers, news and style advice, photographs and film clips from the latest collections, and suggestions about how to create an outfit, all with links to buy the items seen.

By the early 21st century, the East was the centre of both mass and luxury fashion. It was manufacturing its own lines, as well as those for much of the rest of the world, and its increasingly wealthy citizens were keen to shop for fashions too. Tom Ford, who had made his name as creative director first of Gucci and then Yves Saint Laurent, felt this marked a fundamental shift in the international balance of fashion. In Dana Thomas's book *Deluxe: How Luxury Lost its Lustre*, he is quoted as commenting that:

> this is the century of emerging markets ... We are finished here in the West – our moment has come and gone. This is all about China and India and Russia. It is the beginning of the reawakening of cultures that have historically worshipped luxury and haven't had it for so long.

However, the globalization of various aspects of the fashion industry has raised ethical issues concerning, on the one hand, the potential exploitation of labour when manufacturing occurs far from the managerial centre of a company, and on the other, concerns about the homogenizing effects of consumer society, with big brands dominating so much of the world.

Chapter 5
Ethics

Formed in America in 1980, People for the Ethical Treatment of Animals (PETA) has grown to become a global pressure group for animal rights. Its campaigns encompass a number of fashion-related issues, as it forces people to confront the uses made of animals to produce, for example, fur and wool. A 2007 campaign showed British pop singer and model Sophie Ellis Bextor clad in an elegant black evening dress. Her face was perfectly made up: scarlet lips, pale skin, and smoky eyes.

This femme fatale styling was then rendered literal: in one hand, she held up the inert body of a fox, its fur flayed to reveal the red gore of its flesh, its head lolling grotesquely to one side. The tagline 'Here's the Rest of Your Fur' reinforced the message of the cruelty that underpins the fur trade. The campaign's aesthetic drew upon a nostalgic, film noir image. However, 1940s cinematic heroines were frequently shown wearing a fox fur stole draped over their shoulders as a signifier of luxury and sexuality. PETA subverted the viewers' expectations to confront them with the deathliness and horror of fur.

Other print and billboard campaigns have used a similar combination of famous faces, familiar imagery, and the shock of juxtapositions that reveal fashion's underside. The organization's

15. PETA uses arresting imagery and shock tactics to reveal the fur trade's cruelty

aim is to force consumers to understand what goes on behind the sensual façade of fashion photography and marketing, and to examine the way clothes are produced and the processes involved. PETA's slogans use the punchy, direct language of advertising to create memorable taglines that will enter the popular vocabulary. Examples have included ironic double entendres that expose the contradictions at the heart of the fur trade: 'Fur is for Animals', 'Bare Skin, not Bear Skin', as well as 'Ink not Mink', which focused on tattoos as an alternative fashionable status symbol.

PETA's focus on skin itself means the connection between the living animals that provide the fur is continually restaged. Its famous 'I'd Rather Go Naked Than Wear Fur' campaign that started in the mid-1990s brought together supermodels and celebrities, who stripped and stood behind a strategically placed placard. These images were styled like fashion shoots. Despite the lack of any clothing, participants were groomed and lit to emphasize their 'natural' beauty. By using models, actors, and singers as 'themselves', a direct link could be made between their cultural status and value, and the status of PETA's campaign. The message was that if these hugely popular professionals rejected fur, then so should the consumer. Lapses, such as Naomi Campbell's late 1990s defection from PETA and subsequent avocation of fur wearing and, indeed, hunting, have done little to diminish the power of its message. In the early 21st century, a new selection of names, including actress Eva Mendes, signed up to the cause. Images included 'Hands off the Buns' featuring naked celebrities carrying white rabbits.

PETA has raised animal rights' profile within the fashion industry. Its members have invaded catwalks, thrown paint, and famously a frozen animal corpse, at those the organization perceived to be responsible for fur's continued place within fashion, and pushed for new regulations on the treatment of sheep in the wool trade. Its activists' work has not just underlined the needless cruelty involved in the fur trade; it has also shown how fur is often misunderstood as a 'natural' product to wear, despite the fact that most fur is farmed and, once obtained from the animal, goes through various chemical treatments to remove the flesh and prepare it to be used as fabric.

Although PETA's aims are admirable, their approach raises further ethical questions. The group's appropriation of the visual language of fashion and, indeed, wider youth culture has led to accusations that it continued to sexualize and exploit women in the name of animal rights. This is a familiar charge: British-based Respect's 1980s campaign 'One Fur Hat, Two Spoilt Bitches' depicted a model with a dead animal stole and was also seen as positioning women as dumb, sexualized objects. This tension problematized the campaign's message. It can be read as another means to grab the viewer's attention, confront her with the thoughtlessness of wearing fur, and shock her into taking notice. However, to do this, it deployed the highly sexualized visual codes that dominate much contemporary advertising. This controversy highlights the contradictory impulses present within such campaigns. While great focus is placed upon one ethical problem, another equally significant moral issue is accepted, and arguably embraced, as the status quo.

The fashion industry's status is ambiguous. It is a hugely profitable international business and source of pleasure to many, yet it also incorporates a range of moral tensions. From the way women are depicted to the way garment workers are treated, fashion has the ability to represent both the best and the worst of its contemporary culture. Thus, while fashion can be deployed to shape and express alternative as well as mainstream identities, it can equally be repressive and cruel. Fashion's love of juxtapositions and exaggeration can frustrate and confuse, or even reinforce, negative practices and stereotypes. Its focus on appearance has led to its continual condemnation as superficial and narcissistic.

Los Angeles-based T-shirt manufacturer American Apparel is another case in point. Its mission statement has, from the company's inception in 1997, sought to move away from outsourced manufacturing to create a 'sweatshop-free' production line. Unlike other brands that focus on basic wardrobe staples, it has refused to have its garments made up in developing countries, where it can be hard to maintain control of workers' rights and factory conditions. American Apparel instead uses local people, and thus contributes to its community. Its shops include in-store exhibitions of locally and nationally known photographers, and its cool, urban basics have become hugely popular internationally. Its advertising campaigns reinforce its ethical credentials and focus on its workers, frequently using its own shop assistants and administrative staff as models.

However, once again the mode of representation used has caused widespread comment. Dov Charney, the owner of American Apparel, favours a photographic style that is akin to snapshots – candid images of young women and men, often semi-clad, their bodies twisted towards the camera. As Jaime Wolf wrote in a *New York Times* article:

> the ads are also highly suggestive, and not just because they are showcasing underwear or clingy knits. They depict young men and women in bed or in the shower; if they are casually lounging on a sofa or sitting on the floor, then their legs happen to be spread; frequently they are wearing a single item of clothing but are otherwise undressed; a couple of the young women appear to be in a heightened state of pleasure. These pictures have a flashbulb-lighted, lo-fi sultriness to them; they look less like ads than photos you'd see posted on someone's Myspace page.

This aesthetic is not new; it draws upon Nan Goldin's and Larry Clark's graphic images of youth culture from the 1970s. Nor is it unusual to see it used within fashion imagery: Calvin Klein has, for decades, used a similar combination of arresting shots of young models to promote simple designs. It permeates style magazines and online social sites, as well as American Apparel's own website, which presents the images as collections to flick through. They therefore used a familiar set of visual codes in their unstaged-looking set-ups and their casual sexuality.

American Apparel's imagery used a fun, sexy aesthetic that might be expected of a youth-orientated company, but which jarred with traditional ideas of the way a 'worthy' company concerned with ethical issues should be presented. As with the anti-fur campaigns, when a product or cause is positioned as ethical, the use of potentially dubious, sexualized imagery is particularly open to be judged. If one aspect of contemporary morality is being addressed, this sharpens awareness of other possible issues contained within every aspect of an organization or brand's output. While the imagery American Apparel uses chimed with its target youth audience's tastes, it simultaneously exploited an amateur porn aesthetic that had come to pervade early 21st-century culture. Since fashion's own moral status is so fraught, and its role in constructing contemporary culture can be so problematic, it is perhaps unsurprising that ethical messages and practices can be perceived to be undermined by communication methods and representational styles.

Identities and transgressions

While ethical issues that relate to how fashion is produced have gained in significance since the later 19th century, it was the ways in which fashion could be used to change someone's appearance that drove earlier commentaries. Moral concerns centred on the ways that fashion can play tricks, enhancing the wearers' beauty or status, and confusing social codes and acceptable ways to dress and behave. Fashion's close connection to the body and garments' ability to disguise flaws, while also adding sensual fabrics' allure to the figure, added to moralists' fears about both the wearers' vanity, and the effect fashionable clothing had on onlookers. Historically, more was written by those who felt fashion implied narcissistic tendencies, pride, and foolishness than by those wishing to praise it. In the 14th century, for example, text and imagery depicted over-emphasis on appearance as sinful, since, for men and women, it signalled a mind focused on surfaces and materialism rather than religious contemplation. Wearers' uses of fashion to create new identities or to subvert conventional expectations about how they should look meant it could challenge social and cultural divisions, and confuse onlookers. Such anxieties have remained central, where transgressions from the norm have potentially brought moral outrage upon fashion and its adherents.

Although respectable women and men were expected to demonstrate awareness of current fashions in their dress, too much attention to detail was open to question. Fashion was also judged as inappropriate to older people and to the lower classes. This did not, however, prevent fashion's spread. In the 17th century, Ben Jonson's play *Epicoene, or The Silent Woman* included comments that reveal some of the key issues that made fashion dubious. In the play, plain women were deemed more virtuous, while beauty was claimed to entrap men. It also chastized older women who sought to follow fashions in dress and beauty. The character Otter asserts that his wife has:

> A most vile face! And yet she spends me forty pounds a year in mercury and hog's bones. All her teeth were made i'the Blackfriars, both her eyebrows i'the Strand, and her hair in Silver Street. Every part of the town owns a piece of her.

The idea that beauty could be bought, in this case including mercury to turn the face fashionably pale, underlined fashion's inherent duplicity. Mrs Otter's shopping trips meant her appearance belonged to fashionable retailers rather than to nature. She was not just tricking her husband, therefore, but foolishly spending money to recapture her youth.

This theme was developed in sermons, pamphlets, treatises, and imagery in subsequent periods. In the late 18th and early 19th centuries, caricaturists, most notably Cruickshank and Rowlandson, showed elderly women transformed by wigs and beauty preparations, their bodies remoulded by padding and hoops that defined the figure and brought it in line with contemporary ideals. In the 1770s, it was towering wigs topped by foot-long feathers that were most mocked; by the following decade, it was the padding added to the back of dresses; and by the turn of the century, thin women were ridiculed for looking even skinnier in newly fashionable column dresses, while plump women were taunted for looking fatter in the same fashions.

Such criticisms reflected attitudes to women, their bodies, and their status in society. While women were certainly viewed as less important than men, moralists policed their clothing, gestures, etiquette, and deportment. Class also played a significant role, with differing standards and expectations for elite and non-elite women. Importantly, all women were expected to uphold a respectable appearance, to distinguish themselves from prostitutes, and avoid bringing shame upon their families. Women therefore needed to think carefully about how they used fashion; too much interest was problematic, but too little interest could also render women dubious. Fashion's role in shaping gender meant that it was a significant element in people's projection of their individual and group identity. Men were far less criticized for their choices, but they still had to maintain their appearance in relation to their class and status. However, younger men who were too interested in fashion did come in for strong moral condemnation. In the early 18th century, *The Spectator* magazine described foppish students as 'vain Things' who, just like women, 'regard one another for their vestments'. This was perhaps the last period when fashionable menswear was already flamboyant in colour, decoration, and style, and therefore greater effort was needed in order to be transgressive. As *The Spectator* indicated, to do so was to challenge expectations, and risk being regarded as feminine.

16. Eighteenth-century Macaronis were mocked for their exaggerated style and self-conscious deportment

Doubts were cast about the sexuality and gender of many such men. In the 1760s and 1770s, Macaronis, like Fops, who were their most direct predecessors, drew ridicule from caricaturists and commentators. Named after Italian pasta, these young men flaunted their associations with the Continent in their brightly coloured clothes. Their clothing exaggerated contemporary fashions and featured oversized wigs, which were sometimes powdered red or blue instead of the more usual white. They wore coats that were cut extra-tight and curved towards their backs, and were often depicted posing in an affected manner. Macaronis thus offended masculine ideals on a number of counts; they were deemed effeminate, unpatriotic, and vain. Various loosely formed groups of overly fashionable young men superseded them, each of whom used dress to flaunt difference and transgress social ideals. These included the Incroyables of the French Revolution and, in the 19th century, Swells and Mashers in England and Dudes in America. In each case, exaggeration, 'foreign' fashions, and close attention to grooming and accessories distinguished their style and brought claims that they were threatening masculine ideals, and therefore the status quo.

From 1841, *Punch* magazine took pleasure in ridiculing fashions, as well as showing women in crinolines, corsets, and bustles contorted into elaborate shapes in the name of fashion. Alongside these satirical comments were more serious complaints from doctors that women risked their health when they wore whale-boning, but these did little to deter the popularity of such garments. Gender continued to be a major issue. Women needed to wear such underwear in order to be perceived as feminine, yet they were accused of irrationality for wearing such restrictive garments. This double bind extended to clothing that could be seen as too masculine, even if it was more practical than high fashion. In the 1880s, when women began to enter white-collar jobs, the so-called tailormades that they wore, based on a male suit but worn with a skirt, were seen as turning women into men. Indeed, as in all these examples, dress was seen as a signifier of the wearer's gender, sexuality, class, and social standing, and any ambiguities could lead to misunderstandings and condemnation.

This is apparent in the lingering idea that women should not wear trousers, which were felt to disrupt gender roles and imply that women aimed to take on men's powerful status. These concerns extended well into the 20th century. In 1942, the number of women wearing trousers whom she saw in Paris appalled actress Arletty. Despite the hardships of the war, she felt there was no excuse for such behaviour, and that:

> It is unforgivable for women who have the means to buy themselves boots and coats to wear trousers. They impress nobody and their lack of dignity simply proves their bad taste.

This not only revealed the horror with which loss of femininity could be perceived, but stressed the social element of such moral judgements. Working-class women in certain occupations, including mining and fishing, had worn trousers or breeches since the 19th century. However, they were effectively invisible – literally, unseen by most people outside their immediate environment, and metaphorically, since the middle classes and the elite did not value them.

Class has been a persistent theme within moral concerns about the ways in which fashion can disguise someone's true status, or indeed flaunt it as defiance against authority. In the 20th century, establishment mistrust of dress that defied middle-class ideals of respectability and decorum was compounded by the rise in the number of deliberately provocative subcultural groups. In early 1940s France, 'Zazous', both male and female, caused consternation with their elaborately detailed suits, sunglasses, and American-inspired hairstyles and cosmetics. Public and media outrage at their fashions brought together a number of familiar issues. Foreign styles were seen as unpatriotic, particularly during wartime restrictions, even if the Americans were Allies. Exaggerated garments and make-up broke class-based notions of good taste, and paraded Hollywood's overblown style of self-presentation. Although their styles remained confined to a small number of youths, Zazous' emulation of film-star fashions and love of jazz music was a visual and aural confrontation with French culture, at a time when it was already under threat from Nazi occupation of the country.

In subsequent decades, youth culture presented a continued disruption to social codes of behaviour and display. In Britain, class played a significant part in shaping subculture's nature. In the 1960s, Mods aped middle-class respectability in neat, sharp suits, while Skinheads toughened up this style to assert a strong working-class identity, based on workwear. In each case, youth style was driven by a combination of its members' search for excitement and devotion to particular music styles. In the early 21st century, a more diffuse group within working, and unemployed, youth emerged. 'Chavs' were condemned as tasteless, for their unselfconscious flaunting of obvious branding and disregard for middle-class ideals of style. Media coverage exposed embedded class prejudice, as the term quickly became associated with criminality amongst teenagers on council estates. Chavs' aggressive sportswear styles were connected to negative stereotypes of the working class, as an easily grasped visible incarnation of inner-city lawlessness.

Media outrage at each new incarnation of youth style demonstrated the impact that such breaches of the status quo had. In Japan, Tokyo's Harajuku area has, since the 1980s, been a focus for street fashions, as young people evolved new ways to wear and combine garments. Teenage girls upset traditional ideals of femininity to create spectacular new styles which freely combined elements from a range of sources, including high fashion, past subcultures, cartoons, and computer games. Indeed, their composite styles mirrored the fantasy self-styling of computer avatars, which are hugely popular in the Far East. Harajuku's street fashions defy parental expectations that girls should present a demure and restrained image. Pop singer Gwen Stefani's creation of a team of four 'Harajuku Girls' dancers, who appear in her videos and live performances, added another layer of controversy to these styles. Korean-American comedian Margaret Cho has criticized Stefani's appropriation of this Asian fashion style and her use of these 'Harajuku Girls' as offensive, and stated that 'a Japanese school uniform is kind of like a blackface'. This suggested that the dancers represented a stereotype of ethnic identity, used to enliven a white performer's show. Stefani's fashion is itself influenced by Japanese street style, but her dancers take this further. They literally embody concerns not just about foreign inspirations in dress, but more seriously, who has the power to make such appropriations, as well as ethical concerns about ethnic stereotyping.

Another very different incarnation of this is the confused and often excessive response to young Muslim women who choose to wear the hijab as a symbol of religious and ethnic identity. Post 9/11 fears of Islam, combined with public and media perceptions of such displays of difference as transgressive, have led to girls being banned from wearing the hijab in some French schools. This has caused outcry, and hardened some Muslim women's belief in the importance of the hijab as a symbol of not just their religion, but also to question Western ideals of femininity and exposure of the body in contemporary fashion.

This issue sharpens the way specific examples of moral outcry concerning the way ethnic minority groups are presented and treated in relation to dress and appearance. The under-representation of non-white women within the modelling world is a major problem within the industry. Despite media protests and one-off editions, such as Italian *Vogue*'s July 2008 edition, which used black models throughout its editorial pages, white women dominate on the catwalk, as well as in fashion photography and advertising. As leading model Jourdan Dunn, who is herself black British, remarked, 'London's not a white city, so why should the catwalks be so white?' Fashion's persistent disregard for diversity is symptomatic of inherent racism within the wider culture.

Representation, in terms of actual models and their images within magazines, requires a shift in attitudes within the fashion industry and a recognition that it is unacceptable to continue to focus on white models.

Regulation and reform

Alongside protests against the ways men, and particularly women, are represented in fashion imagery, there have been various attempts to control or manage the ways in which fashion is produced and consumed. During the Renaissance, sumptuary laws continued to be imposed to try to maintain class distinctions, by limiting certain fabrics or types of decoration to particular groups, or to impose ideals of modesty on the population. For example, in Italy, legislation was passed that sought to regulate attire worn for rituals such as weddings, as well as to limit the amount of décolletage women of different classes were permitted to display. Such laws were regularly instigated across Europe, although they had limited success, since they were difficult to police. As Catherine Kovesi Killerby has written in relation to Italian laws that expressed social concern about excessive display in dress, 'by their very nature, [they] are self-defeating: to curb luxury by the outlawing of one form that luxury happens to be taking itself generates new forms as the way to avoid persecution'. Since fashion continually mutates, albeit at a slower rate during this early period, it is hard for the legislature to keep up with these changes, and as Killerby notes, wearers are equally inventive, changing styles to dodge laws and create new incarnations of a style.

Sumptuary laws declined during the 17th century, although they were resurrected with greater success during the Second World War. While earlier periods had seen bans imposed on importation of foreign goods for economic and nationalistic reasons, the length and extent of this war meant any such laws were compounded by severe restrictions on international trade due to widespread sea and air warfare. Shortages led to rationing in many of the countries involved. In 1941, Britain regulated production and consumption of clothing, by issuing coupons that could be exchanged for garments throughout the year. The number of coupons issued to each person changed over the course of the war and post-war period, but imposed a serious limit on access to clothes. Regulations in Britain, America, and France also stipulated how much fabric could be used in clothing production, and stripped back the amount of decoration that could be applied. This stark shift in access to fashion was tempered by the British Utility scheme that employed well-known fashion designers, including Hardy Amies, to design outfits that followed the legal limitations while remaining stylish. The lack of new clothes meant it was hard to circumvent wartime restrictions, though, and public and media attitudes hardened towards excess, which was seen as unpatriotic and against the war effort.

After the war, Soviet bloc countries were able to continue this limit on fashions and attempted, with varying degrees of success, to condemn fashion as anti-socialist. In East Germany, Judd Stitzel writes that:

> officials sought to channel and control female desire by connecting women's rights as consumers with their roles as producers and by promoting rational 'socialist consumer habits' as an important component of citizenship.

Work-inspired garments including aprons and overalls had limited appeal, however, and, as in other socialist countries, including Czechoslovakia, an uneasy coalition of state-sanctioned fashions and fashion imagery was developed alongside more functional styles. These attempts to reform fashion and strive for a more ethical form of dress harked to 19th-century dress reformers such as Dr Gustav Jaeger who had encouraged men and women to reject fashion's excess and adopt natural-fibre clothes, and feminists in

Europe, Scandinavia, and America who called for greater equality and rationality in clothing.

Late 20th- and early 21st-century versions of these impulses to regulate and create clothing that does not harm animals, people, or the environment have begun to make inroads into mainstream as well as niche fashion. Spurred on by the Hippies and connected movements in the 1960s and 1970s towards more natural fashions and concern for ethical issues, at the turn of the 21st century designers as well as bigger brands tried to reconcile developments in consumerism with the need for more thoughtful design and production practices. Since the early 20th century, moves were made to regulate wages and conditions for workers. This was prompted by disasters such as the Triangle Shirtwaist Factory fire in New York in 1911, when 146 immigrant workers were killed. The factory contained an unknown number of subcontracted, poorly paid workers in an overcrowded, cramped environment, which meant that many could not escape the blaze that broke out on the top floors. Although such incidents brought widespread protests against sweatshops and calls for a minimum wage, these practices still have not been eliminated. As rents rose in major cities, mass production moved further out, and eventually migrated to poorer countries in South America and the Far East, where labour and property was cheap. So-called 'Fast Fashion', where brands strive to provide the latest fashions as soon as they have been seen on the catwalk, has led to strong competition to introduce new styles throughout the year, at the cheapest prices possible.

Popular high-street names have been accused of using suppliers that rely on child labour. In October 2008, a report by the BBC and *The Observer* alleged that three of low-cost brand Primark's suppliers used young Sri Lankan children from refugee camps in India to sew decoration onto T-shirts, in appalling conditions. Primark sacked these suppliers as soon as it was made aware of the situation, but the report suggested that there was a problem at the heart of the contemporary fashion industry. Cheap clothing's easy availability democratized access to fashion, but also encouraged consumers to view garments as short term and throwaway, and, combined with fierce competition to produce the cheapest lines, makes exploitation a potential consequence. Mass-market fashion chains have stated that their huge sales volume meant that their clothes could be inexpensive. However, there can be an ethical cost to this approach, as well as a human cost, as supply chains become increasingly diffuse and difficult to track. Journalist Dan McDougall has stated that:

> in the UK the term 'rush to the bottom' was coined to describe the practice of international retailers employing developing world contractors, who cut corners to keep margins down and profits up for western paymasters.

Primark is not the only chain store to face criticism; others, including American-based Gap, have also had problems with their suppliers. Labels such as People Tree in Britain have therefore sought to distance themselves from this approach, and have established close ties with their suppliers, to seek to create sustainable production patterns that can benefit local communities in the countries where their clothes are made. Bigger brands including American Apparel have taken action to prevent sweatshops by using local employees. Both brands have also worked to use fabrics with a low impact on the environment. The poisonous bleaching and dying processes used in denim and cotton production have prompted organic and unbleached ranges to emerge at all levels of the market. What distinguished the clothes produced from earlier ranges in previous decades was manufacturers' recognition that consumers expect fashionable design values even from ethical goods. Smaller labels such as Ruby London, which included a selection of fashionably skinny-cut organic cotton jeans in its range, and Ekovarnhuset in Sweden, which sells its own line as well as other eco-fashion labels, have created clothes that are fashionable as well as environmentally conscious. Even big brands including H&M, New Look, and Marks and Spencer introduced organic cotton lines. High fashion incorporated a growing number of ethical labels too. Stella McCartney refused to use fur or leather, while Danish designers Noir combined cutting-edge fashion style with a strong ethical company policy that included support for the development of ecologically sound textiles.

Other designers promoted the idea of buying less, but investing in more expensive pieces that would last longer. This 'Slow Fashion' encompassed ranges such as Martin Margiela's 'Artisanal' line of handmade garments. *The New York Times*' Armand Limnander broke down the relative cost of these luxury designs to calculate that, for example, a Raf Simons at Jil Sander made-to-order man's suit at $6,000, which took 22 hours to make, was therefore priced at $272.73 per hour. While this did not estimate the cost per wear, it advocated a shift in attitude that rejected quick turnover of styles and seasonal purchases of the latest trend. Not everyone, though, can afford the initial investments needed. However, Slow Fashion identifies one of the core issues within making fashion ethical: that consumption itself is the problem. Fashion's environmental impact spans a wide range of issues from production methods and the practices involved in growing natural fibres such as cotton, to mass consumerism and the public's desire for new fashions.

Japanese chain Muji's recycled yarn knit range suggested one solution; Paris-based Malian designer XULY Bët's designs made from reused old clothes another. These clothes rely upon second-hand textiles and garments, and can be seen in conjunction with the shift towards vintage and flea-market fashion shopping since the late 20th century. These fashions have less impact on the environment and reduce the production process, but they are unlikely completely to replace the existing fashion industry, especially given its huge international reach and the amount of finance tied up in its production and promotion.

There is also a danger that ethical shopping itself becomes a trend. As a global economic downturn set in during the first decade of the 21st century, reports questioned the idea of 'recession chic' and 'feelgood consumerism', based on people's sense of virtue when they bought organic and ethically produced clothes, even if their purchase was actually unnecessary. The question remained whether consumers were willing to own less and to rely less on shopping as a source of leisure and pleasure, and whether ethical brands can assert a new approach to assessing what to buy and remain viable businesses.

Counterfeit markets across the globe which sell copies of the latest 'It' bags demonstrate the continued allure of status symbols, and fashion's ability to seduce consumers' eager for an object associated with luxury and elite style. As fashion's reach has spread across the social spectrum and incorporated internationally known brands, it has become increasingly difficult to police its production or

17. Markets in the Far East sell counterfeit versions of the latest luxury brand 'It' bags at a fraction of their retail price

regulate its consumption. This could only be achieved by a major realignment of social and cultural values, and a change in approach from a global industry that had grown up over centuries to lure customers and satiate their desire for the tactile and visual allure of clothing.

Chapter 6
Globalization

Manish Arora's autumn/winter 2008–9 collection was shown against the backdrop of artist Subodh Gupta's installation of neatly arrayed stainless-steel cookware. This metallic scenography provided an ironic comment on clichés of Indian culture. Gupta's glittering display was also a foretaste of the hard silver and gold tones that dominated Arora's show. His models were dressed as futuristic warrior women. He used a mix of historical references to create gleaming breastplates, stiff mini skirts, and articulated leg pieces. Roman gladiators, medieval knights, and samurai were all evoked, with spiny silver facemasks to reinforce the image of power. These international inspirations were taken further in Arora's trademark use of vividly coloured three-dimensional embroideries, beadwork, and appliqué. These added to the combination of old and new, in their display of traditional Indian craftsmanship that used glittering Swarovski crystals to enhance the effect.

Arora's collaborators were equally diverse. Japanese artist Keiichi Tanaami contributed his psychedelic imagery of huge-eyed children and fantastical beasts as templates for the decoration applied to dresses and coats. Walt Disney's Goofy, Mickey and Minnie Mouse were re-imagined in armour and helmets on a series of garments. The result was a collection that underlined

18. Manish Arora's 2008–9 collection included warrior-woman imagery and embroideries of Walt Disney characters

Arora's ability to produce a coherent look from seemingly unconnected influences and ideas, as well as to reinforce his status as a global designer, able to erase stark definitions of East and West in his elaborate designs. Since he set up his label in 1997, Arora has produced imaginative work that incorporates traditional embroideries and other decorative techniques with Pop Art style colourings and myriad reference points. This embellishment spoke of luxury and excess, and catalogued in minute detail his progress within the fashion industry. During his time showing at London Fashion Week, city panoramas of the Houses of Parliament and the Trooping of the Colour crowded onto full skirts – then, while showing in Paris, the Eiffel Tower appeared. From the start, he aimed to establish a global luxury brand which catered to the tastes of both Indian and international audiences. Indeed, his style rendered these distinctions ever more anachronistic. In most cases, there was no difference between them, and, as Lisa Armstrong noted, Arora 'doesn't seem to be pandering to foreign markets – or attempting to dampen his exuberance'.

The early 21st century saw a steadily growing schedule of fashion weeks across the globe, instant dissemination of trends via the Internet, and financial and industrial growth in countries such as India and China. Arora's own success was a product of India's developing confidence as a fashion centre. It had a long-established reputation for its textiles and craft skills, but it was not until the late 1980s that it began to construct the infrastructure necessary to build a fashion industry. Couture designers began to emerge, and colleges, including the National Institute of Fashion Technology in New Delhi, where Arora studied, educated a new breed of designer. In 1998, the Fashion Design Council of India was set up to promote Indian designers and seek sponsorship. This made it possible for ready-to-wear labels to evolve, and thus created the basis for a broader-based fashion industry with further reach beyond India. Arora's entrepreneurial ability enabled him to gain worldwide publicity, and lucrative design connections. For example, he has produced a range of shoes for Reebok, a limited-edition watches line for Swatch, and a cosmetics collection for MAC that displayed his signature neon-bright colours and love of shimmering surfaces. Business deals such as these provided the platform for Arora to expand his brand.

However, his success should not just be judged by his recognition within the West. Rather, as part of a developing breed of non-Western designers able to command international sales and attention, Arora represented a gradual shift away from the West as the fashion world's core. This process is by no means complete; it is notable that Arora showed in London and Paris to raise his profile with international press and buyers, while still showing in India. The rise of the middle and upper classes in India, though, meant that he and his peers had a considerable potential domestic market, as is the case in other countries that have invested in fashion, including China.

Western fashion cities also benefited from the cachet of including international designers in their programme. London Fashion Week had struggled to maintain its profile and to encourage foreign media and the all-important store buyers to attend its shows. In February 2005, journalists Caroline Asome and Alan Hamilton described how names such as Arora, along with Japan-based Danish-Yugoslavian-Chinese duo Aganovitch and Yung, added interest and diversity to its schedule. These international designers showed alongside London-based Nigerian Duro Olowu, Serbian Roksanda Ilincic, and Andrew Gn from Singapore. Such global names within one city underlined fashion's international scope, and suggested that while national and local styles may in the past have been useful to market designers as a group, these distinctions were less meaningful as a wider range of fashion cities emerged and designers were, subject to financial backing, able to show their collections in any number of sites. Fashion's geography had begun to shift, but as Sumati Nagrath noted, 'since the Indian fashion industry [for example] is a relatively new entrant on the global fashion scene, it has meant that in order to participate in it, the "local" industry has perforce had to operate within a pre-existing system'. However, as other regions evolve, and movement of goods and labour alter patterns of production, the fashion infrastructure that crystallized during the late 19th century may itself begin to alter its focus.

Paris consolidated its position at the centre of Western fashion at this time, but even by the early 20th century, the French fashion industry was concerned about superior business practice in the United States. Once American ready-to-wear developed its own signature during the Second World War, it became possible for ready-to-wear, rather than just couture, to generate fashions. As post-war reconstruction drew upon the American model, and more relaxed styles including jeans and sportswear were marketed internationally, a fundamental shift occurred in fashion even though Paris still wielded considerable influence. Perhaps in the early 21st century a similar process was in train, and this was not necessarily a completely new development. In fact, it represented, at least in the case of India and China, the resurrection of luxury and visual display in dress in countries that had a long history of skills in these areas that had been interrupted by colonialism, politics, and war.

Trade and dissemination

Trade routes had transported textiles across the world since the 1st century BC, linking the Far and Middle East to European cities that dealt in rich textiles. Italy was a gateway between East and West, and had established itself at the heart of the luxury trade in textiles. Northern Europe developed centres for wool production, and Italy was famous for its multi-coloured designs in rich silks, velvets, and brocades. Cities including Venice and Florence produced the bulk of Europe's fine textiles, and its fabrics sometimes bore the imprint of the Mediterranean trade that helped to create them, with Islamic, Hebrew, and

19. Renaissance textiles often combined motifs from Europe, and the Middle and Far East

Eastern texts and designs combined with Western motifs. These cross-cultural reference points were a natural result of trade, which developed during the Renaissance, as nations sought to control particular zones or find new land. During the 15th and 16th centuries, trade grew between a wider range of European countries, and links were made between Portugal, Syria, Turkey, and India and South East Asia, and between Spain and the Americas.

In the early 17th century, first England and then Holland established East India Companies (EIC) that formalized and organized their trade with India and the Far East. Initially, as John Styles has noted, the English EIC was most interested in exporting wool to Asia, and only brought back tiny amounts of very luxurious Eastern textiles, as their designs had limited appeal in England. However, in the second half of the 17th century, the EIC sent patterns, and later samples, to its Indian agents, which encouraged production of patterns based upon an English idea of the 'exotic'. These became very popular, and meant that Western fashion, which drew upon such materials for its impact, incorporated larger amounts of Eastern products. Europe had developed sophisticated maritime knowledge and transport methods to enable this trade, and exploited the innovation, flexibility, and skill of Asian craftspeople. They produced a diverse range of materials, and responded quickly to customer tastes. This produced fertile ground for cross-cultural interchanges and produced designs that merged references from various countries and ethnicities. However, Western taste dominated, and shaped the ways that Asian motifs were used. Consumers were encouraged to appreciate styles from far-flung countries, as reconfigured by EIC representatives who were aware of their tastes and aspirations. The global textiles trade was driven by luxury fabrics' appeal to the senses and Western interest in an emerging idea of exoticism, and was underscored by its considerable money-making potential. This was based upon the elite's desire for extravagant display, something that was common to all countries.

Dress styles tended to remain distinct, despite specific types of garments making the transition from East to West. This included kaftan-like dressing and wrapping gowns worn by European men and women for informal occasions at home, and a parallel fashion for turbans that was well established by the end of the 17th century. Portraits of the period show Western men relaxing in shot-silk wrapping gowns, with turbans covering their shorn heads, as a welcome escape from the powdered wigs they wore in public.

Indeed, Peter Stallybrass and Ann Rosalind Jones have argued that 17th-century identities were less tied to ideas of nation or continent. They analysed Van Dyck's portrait of Robert Shirley, English ambassador to Persia from 1622, to show how membership of the elite was far more central to identity at this time. Shirley is shown in Persian dress appropriate to his social rank and professional status. The lush embroideries of his garments, with polychrome silks on golden ground, demonstrate how much more developed such skills were in the East, and the sumptuousness of Persian attire. Stallybrass and Jones suggest that Shirley would not have perceived himself as European, since this region had no coherent identity at the time. Nor would he have assumed superiority due to his Westerness. He would, they argue, have easily adopted Persian dress as a marker of his new position and as a signal of his deferential relationship to the Shah. Fashionable identities were equally connected to ideas of class and status, but they also connected to regional or court ideals of taste and individual ability to adopt and interpret current trends. However, as Shirley's portrait shows, this identity could incorporate elements of other ethnic expectations for particular social or professional occasions, and, importantly, when living or travelling abroad. The vogue for Turkish-inspired loosely wrapped dresses amongst European women during the following century is further evidence of this, as are the adaptations of real Turkish garments by female travellers such as Lady Mary Wortley-Montague.

Indeed, it would seem that during the 17th century ideals of luxury and display were common to Eastern and Western noble and court circles. Carlo Marco Belfanti has shown that fashions developed in India, China, and Japan during the 17th and 18th centuries, with particular tastes and cycles of styles becoming popular. In Mughal India, for example, tailoring was experimented with, a love of excess permeated design, and fashions in styles of turbans and head wraps emerged. Fashions in cut and design of clothing were also present amongst clerical workers in bigger cities. However,

Belfanti argues that while fashion itself evolved in both East and West simultaneously, it did not become a social institution in the East, and proscribed forms of dress became the norm by the 19th century.

Cross-cultural references spread beyond the elite, though, and represented global influences based upon trade, but reliant upon designs that engaged audiences in the East and West. The West developed its own interpretations of designs from the East. In the mid-18th century, *chinoiserie* decorative styles had swept Europe. Aileen Ribeiro describes these re-imaginings of the East, which prompted textiles covered in pagodas and stylized florals, amongst other reinvented Chinese motifs. This trend can be seen as part of an aristocratic love of dressing up, in this case in a fantastical version of other ethnic and cultural styles. China became a popular theme for masquerades, and the Swedish royal family even dressed the future King Gustav III in Chinese robes while at its summer palace in Drottningholm.

Chinoiserie was a fashion that resulted from fanciful Western interpretations of Eastern design. However, the huge popularity of Indian chintzes during the 18th century showed the impact that Indian fabric manufacture and print design could have upon a market that extended well beyond Europe to include colonies such as those in South America. The cheapness of many Indian cottons meant they were within the reach of a far wider population than ever before. This also meant that international tastes in textile design and type, as well as access to fashion, and easy-to-wash clothes were within the reach of all but the poorest. In fact, in the 1780s the so-called 'calico craze' caused consternation amongst governments, who feared their indigenous textile trades would be made redundant. Sumptuary legislation was passed in various countries, including Switzerland and Spain, while in Mexico Marta A. Vicente writes that women reportedly sold their bodies to buy these foreign fashions. Ultimately, though, what Western countries discovered from this quickly spreading fashion was that rather than fighting its popularity, they should use it to build up their own textile industries, and apply what they could learn from Indian textile producers to profit from the craze, as was the case in Barcelona, for example.

This was part of what would become a significant global shift from the innovative and adaptable Indian textile trade towards the increasingly industrially led West, which would gain pace during the 19th century. As England in particular developed a succession of inventions designed to speed up textile manufacture, it overtook Indian textile production, and this led to the almost complete abandonment of trade in hand-woven Indian textiles by the 1820s. Fashion had shifted its balance of power in terms of textile production as Western countries began to rely far more upon their own manufacture and export of fabric, rather than imported cottons. The Western fashion system quickly emerged in the form that would dominate for the coming century and beyond. Mechanization enabled European, and later American, textile mills to respond rapidly to tastes and fashions. In the 1850s, European inventions of synthetic dyes, notably William Perkin's discovery of vivid mauve aniline colours, all but wiped out the natural dyes industry in other parts of the world. Sandra Niessen has noted that this led to these new, vibrant hues spreading across the globe, which altered the look of traditional as well as fashionable dress everywhere from France to Guatemala.

The build-up of Western-owned colonies over the course of the 19th century saw the exploitation of textile trades in the hands of European powers. Despite racist attitudes apparent within Victorian culture, both elite and middle-class consumers continued to admire non-European products. This included Indian textiles and Japanese dress. Arthur Lasenby Liberty's department store on Regent Street in London was established in 1875. It sold furniture and decorative items from the East, as well as clothing and textiles inspired by the owner's admiration for looser, more softly coloured Asian designs and the draped gowns of

medieval Europe. However, Tomoko Sato and Toshio Watanabe
have shown that Liberty's attitudes to the East were conflicted, and
expressed the vexed relationship between Western exoticized ideas
and the reality of Asia. In 1889, he went to Japan for three months,
and, like other contemporary commentators, was pleased to see
that silks had become thinner and easier to handle under Western
influence, but did not approve of changes in colour and design that
had also occurred. Once Japan had reopened to the West in the
1850s, and began to modernize, both men and women began to
wear Westernized dress, as well as traditional styles. For Victorians
such as Liberty, this change disrupted their view of the East. This
ideal was complex, as it had evolved over time, shaped by Western
perceptions of difference, and reinterpretations of Eastern design
that responded to the Orient as the opposite of industrialized
Western countries. While the late 19th-century cult of Japan
tended to see the East as static, in contrast to Western fashion's
swiftly changing styles, Japan itself was quickly absorbing Western
influence to reconfigure its own designs.

Local and global

At the start of the 20th century, the fashion industry had therefore
evolved from this complex history. While on the one hand, certain
countries, especially those under the generalized Western idea of
the East, were seen as a rich and sensual source of inspiration, on
the other Westerners tended to view the rest of the world as a
resource rather than as equals. Trade networks had shifted and
transformed over the centuries, but tended to be controlled by
Western powers. The fashion industry had global trade links, but it
was yet to become globalized, with corporations that were truly
international, and fully fledged fashion systems in multiple
countries across the world. This is not to say that fashion did not
exist outside the West; style changes emerged on other continents,
fuelled by local tastes and social structures. However, cyclical
fashions generated by designers, manufacturers, and promoted by

retailers and media were to evolve in the second half of the 20th century.

During the interwar period, French haute couture was very powerful and drove international trends. However, its success was predicated on sales not just of individually made garments, but also of designs that manufacturers in other countries could buy and reproduce. At the same time, cities such as London and New York sought to establish their own fashion identities, with increased focus on designer names and fashion-led manufacturing. This process laid the foundations for the fashion industry's post-war acceleration and growth. High fashion was still enthralled by French style, but other countries were fast developing their own marketable fashion signature, particularly in terms of ready-to-wear. America is a case in point: in the 1930s and 1940s, its fashions were frequently promoted in relation to patriotic myths of a coherent national identity. By the early 1950s, although it continued to use emblems of Americanness in its design and imagery, it was more concerned to promote its international fashion signature and status. This is illustrated by American *Vogue*, which increasingly covered a wider range of countries' fashion collections during the 1950s. Alongside Paris and London, which had long featured in its editorial and advertising pages, collections from Dublin, Rome, and Madrid were covered each season. Even though *Vogue*'s focus remained European and Western, this showed how aspirations towards high fashion status had spread.

As these cities began to emerge as style centres, America built on its strengths in simple, easy-to-wear separates and neat dresses. These were sold to wider markets in the post-war period, but most importantly, denim jeans and sportswear came to dominate the global scene after the war. Worn by all ages, genders, ethnicities, and classes, denim was the most significant factor in the globalization of a recognizable style statement. Although jeans are not necessarily automatically fashion, their rise in status expressed consumer desire for clothing that could be worn with a range of formal and informal garments and could be adapted to fit with individual style. By the early 21st century, jeans represented a huge international market, and although this could be read as a homogenization of fashion and therefore of global visual identity, denim is diverse and can in fact expose national, regional, subcultural, and individual identity through its myriad permutations. In Brazil, for example, Mamao Verde produced skintight denim jeans with sparkling decoration to emphasize the wearer's curves. In Japan, denim was fetishized, and collectors sought out rare pairs of vintage Levis, as well as indigenous brands such as Evisu, which included baggy-cut jeans decorated with its signature logo print. It is not just designer and sought-after brands that provided denim with its diversity. Individuals created their own distinct denim, as the indigo dye becomes fainter through washing and creases in line with the wearer's body. Jeans are frequently customized, or worn with a mixture of second-hand and new clothes to create micro fashions particular to a specific area. In this way, homogenization and globalization could be resisted or at least given a different feel in relation to local rather than international impulses, and through the wearer's creativity.

Wearers' individualization of their clothing and accessories can thus complicate simple readings of globalization's impact on visual style. However, in many cases big brands' spread across the globe can lead to high streets, shopping malls, and airport duty-free lounges all too often comprising the same familiar labels. The quick response of chains such as Zara to local fashions spotted on the streets and integrated into their designs can lead to differences in what is sold in their branches in different countries and even cities. However, in other cases, Western brands' dominance of the marketplace can lead to visual similarities between fashion styles amongst particular social classes internationally, as was the case amongst the elite in earlier periods. The same brands of sunglasses, handbags, and other accessories are shown in international fashion magazines, and bought by consumers who wish to attain what might be called this global high-fashion style. Its precursor is clearly Parisian couture's domination since the 17th century, but by the time the international jetset of the 1970s had emerged, it might just as easily be an Italian or American label that was coveted. The wealthy in many cities adhere to their own version of this style, to produce transnational fashions that rely more on social than geographical boundaries.

However, nuances still emerge, in relation to national ideals of beauty and gender, for example. Age is another important factor that shapes how such fashions are interpreted. In the 1990s, British brand Burberry's signature scarves, trench coats, and handbags became popular amongst South Korean youth. While this can be seen as an example of homogenization, the brand's signature check was worn in a different way. In Korea, as in Japan, a complete designer outfit was aspired to, with everything from shoes to hairgrips heavily branded. This conspicuous consumption was not fashionable in the West, where emphasis was placed on a wearer's ability to combine labels and mix them with vintage or non-branded goods, and logos were only periodically fashionable. Young South Koreans therefore subverted Burberry's brand image of restrained British upper-class taste by their enthusiasm for its goods.

Margaret Maynard has identified this complex interplay between increased international merging of fashion trends as a result in part of global brands, as a product of changes in the late 20th century. She argues that this marks the moment that globalization began to impact economic, political, and social life, therefore affecting the fashion industry. She cites international events including the collapse of communism, demise of postcolonial rule, growth of multinational corporations and banking, and world media and Internet, as responsible for greater dissemination and circulation of fashion garments and imagery, and fashion markets awakening in myriad countries. The growth in international travel and immigration patterns has speeded up the breakdown of boundaries and the concomitant growth of globalization. This process has led to ethical issues in terms of, for example, Western capitalism's search for cheaper manufacturing, and the parallel rise of Fast Fashion has seen its own industrial production decline. Brands from luxury names such as Gucci to mass-market Gap have outsourced their manufacturing to countries including China, Vietnam, and the Philippines. This has led to globalization's most pernicious effect – workers' exploitation. It has become difficult to track suppliers and maintain factory standards. Workers are abused and underpaid, and are frequently drawn from the most vulnerable sections of the population, for example children or recent immigrants. Globalization has therefore provided a mask behind which unjust industrial manufacturing practices can hide. The fashion industries' vast geographical scope has made it all too easy for non-unionized labour to be used to provide cheap fashions for the growing international market. It has also meant the luxury conglomerates, notably LMVH, now dominate the industry, alongside companies including sportswear-based and youth-orientated labels such as Diesel and Nike. However, Maynard contends that local differences are still able to break through this potentially homogenized mass of globally available goods, and therefore a completely uniform look or idea of fashion has not been universally imposed.

Senegal's fashions are an example of this locally formed popular culture, which appropriates from, but can equally resist, the mass culture of fashion produced by huge corporations. Senegalese youth look to diverse global influences in their style, and confidently integrate European and Islamic elements, as well as different types of fashion. While jeans and African-American trends are apparent, young people also commission more formal designs from local tailors. Hudita Nina Mustafa has shown how important self-presentation has been in Senegal since well before the French colonial period. She describes how men and women wear hybrid Eurafrican fashions, as well as garments that are specific to their region. The capital Dakar's highly flexible tailors, dressmakers, and designers, including Oumou Sy, who also exports her work to Tunisia, Switzerland, and France, encapsulate this sophisticated, cosmopolitan use of fashion. They create garments that are inspired by current local styles, traditional forms of dyeing and decoration, international celebrities, and French couture. Global networks of trade enable Senegalese traders to commission fabric designs from Northern Europe, buy textiles in Nigeria, and trade in Europe, America, and the Middle East. The country's fashion system therefore integrates local and global impulses to create fashions that connect to consumers. It is at once part of the globalized fashion industry, yet retains its own commercial patterns and aesthetic tastes. Dakar's vibrancy as a fashion capital exemplifies the ways in which fashion industries can coexist and overlap in the 21st century. Indeed, as Leslie W. Rabine has suggested, Africa as a whole incorporates a variety of fashion and entrepreneurial types that work both within and on the edges of the Western capitalist industry, 'through such networks, peopled by suitcase vendors who transport their goods with them in suitcases and trunks, producers and consumers create transnational popular culture forms'. Thus, street traders, like the pedlars of earlier periods, as well as travellers and tourists, and long-term and permanent immigrants spread fashion garments and accessories across the globe. These formal and informal methods blur clear-cut distinctions of national identity, just as the spread of global branded goods do. In fact, combined with the international trade in second-hand clothes, they help to resist the homogenized ideal that such brands all too often represent.

High-fashion collections shown in European and other cities also incorporate notions of transnational designs, which fuse references from a wide range of cultures and ethnicities, without being clearly defined by any one geographical region. Manish Arora's work is an example of this, since he combines East and West in terms of both design and decoration. Unlike early 20th-century designers such as Paul Poiret, who used Middle- and Far-Eastern influences from the perspective of a colonial Westerner, Arora refuses such hierarchies. However, the 'Orientalizing' influence of the West is deeply embedded in visual and material culture. Questions remain about who produces, controls, and dominates the use of images and fashion styles. Cultural appropriation abounds in fashion, and provides a rich palette of cross-fertilization in ideas, styles, and colours. However, José Teunissen has asked:

> the image exotic cultures have of themselves is often determined by the dominant West. What is Indian after all? Is it what people of India call Indian, or what we in the West, with our colonial past – once labelled Indian?

In the early 21st century, this has remained a fraught issue, as has the question of whether it is different for a Western designer to use 'exotic' references, given the long, and hugely problematic, histories of colonial rule and domination. Postmodernism may have provided justification for designers' playful cross-fertilization of ideas from a wide range of ethnic and historical reference points, as seen in John Galliano's work, for example. However, it cannot completely erase contexts in which the fashion industry evolved or the historical meanings of such appropriations to enable an equal exchange, either in design and aesthetic terms, or in other aspects of the industry such as trade. As more and more countries begin to promote their fashions internationally, these differences will perhaps diminish. This process will not be complete until a sufficient number of designers, luxury brands, and ready-to-wear manufacturers from non-Western countries have the same power and reach as LVMH and its peers.

Fashion weeks, which group together a particular country's or city's designers to show their seasonal collections, continue to provide a focus from which to promote an area's visual identity as well as to develop and provide a platform for its fashion designers. Fashion is a huge industry with great economic and cultural significance, and the spread of fashion weeks in various South American cities, for example, shows how they can begin to

20. Issey Miyake's angular pleating, from 1990

establish alternative fashion centres. The impact that non-Western designers can have on the globalized market was demonstrated by the huge success of Japanese designers who showed in Paris from the late 1970s and early 1980s. At this time, it was still necessary for designers to show within an established fashion week to gain sufficient publicity and exposure to international store buyers.

Japanese designers such as Yohji Yamamoto, Rei Kawakubo of Comme des Garçons, Kenzo, and Issey Miyake's work shocked the Western fashion world into the realization that high fashion could emanate from beyond its confines. Importantly, Japanese fashion also provided an alternative vision of body and fabric and the dynamic between them.

Issey Miyake, for example, produced clothes that overturned Western ideals of beauty and form and presented tightly pleated textiles sculpted into points that pulled out from the figure. He recreated femininity in line with architectural notions of space, rather than cutting fabric in towards the natural form. His clothes often swept upwards, and jutted out to emphasize the contrast between body and garment. His work is carried out on the international stage, shown and sold in cities across the world. However, in the 1990s, Miyake commented that despite, or perhaps because, global 'boundaries are being destroyed or re-defined before our eyes, daily... I think they are necessary. After all, boundaries are the expression of culture and history'. His desire to maintain his Japanese identity, yet simultaneously to produce designs with international resonance and appeal, is at the heart of questions about the fashion industry's globalization. Trade networks, production, consumption, and design have all increasingly become tied to globalized fashion systems since the late 20th century. Globalized fashion has not completely repressed either local or individual expression through fashion, either for designers or for wearers. However, the recession that developed in the early 21st century may accelerate the development of non-Western fashion design, built upon already well-established production patterns, and produce a dramatic shift in the balance of fashion power.

Conclusion

Teri Agins' significant 1999 book *The End of Fashion* described what she saw as the industry's shift from fashion to clothing in the late 20th century. She argued that French couture had been slow to realize the need to focus on wearable classics at affordable prices, and was surviving on its franchises, in particular its worldwide perfume sales. At the same time, big European corporations had realized that American designers such as Michael Kors at Celine could bring in more sales for their stable of brands than the more dramatic British designers such as John Galliano at Dior. Agins outlined designers' focus on innovation in marketing rather than fashion design. Concomitant with this was the public's exhaustion with fashion, and increased interest in high-street chains including Gap and Banana Republic, which were reliable for wardrobe staples, as well as occasionally setting fashions. Agins' argument was convincing, and came at the end of a decade that had seen international recession and market crashes in the Far East. As she noted, since minimalist designs had been fashionable, pared-down dressing was itself part of a trend away from elaborate fashions.

So, did fashion end in the 1990s? Was this the triumph of clothing? Agins certainly showed an important trend in the international market. However, what is perhaps most interesting is that it was a trend. As she said herself, minimalism was a fashion at the time, and thus its presence at all levels of the market was itself part of this fashion. It is important to note that other trends were also apparent. Young designers such as Alexander McQueen began successful careers in the early 1990s, and built labels predicated not just on franchising but on innovation in design. Significantly, the mid-1990s, the moment Agins identified as the turning point away from fashion-led clothing, was also the stage when a growth in interest in craft skills and detail began to emerge in designers including Matthew Williamson's work. Perhaps what Agins identified was not the end of fashion, but rather an example of its flexible and constantly mutating form. As cultural, social, and economic contexts change, so too do designers' inspirations, and consumers' needs and, more importantly, desires.

Certainly, there was a major trend towards workwear-inspired fashions at street and high fashion levels, which encompassed everything from cargo pants to grunge, and stark, intellectual minimalism from designers such as Jil Sander. However, it is important to remember that various fashions exist simultaneously; there was also a revival of goth fashions, and dark and fetishistic styles in high fashion. Alongside this were Williamson's fruit-coloured fashions, which brought luxury details and vibrant prints back into focus. While America still favoured Gap, in Europe it declined, its loose fit and anonymous style unable to compete with the rise of Topshop in Britain and Kookai in France, for example, as fashionable and exciting alternatives. Agins therefore wrote of a tipping point in American-focused fashion and clothing tastes and lifestyle, just as alternatives to this vision were beginning to take hold of the public's imagination. She was therefore right to identify the importance of this moment in fashion history, but fashion's apparent demise was actually the moment before its revival as a driving force from high street to couture level.

What Agins' work did was remind us of fashion's inherent ability to absorb outside influences, and to reinvent itself in line with, and sometimes even in anticipation of, new lifestyles and tastes. By the early 21st century, clothing was still an important part of the market, as it always had been, and, as Agins stated, wardrobe classics were as needed as ever. However, new couturiers, such as Alber Elbaz at Lanvin, Nicolas Ghesquiere at Balenciaga, Stefano Pilati at Yves Saint Laurent, and Christoph Decarnin at Balmain, caused international excitement about French fashion once again. Even if most could only aspire to buy their iconic handbags, their seasonal style statements were quickly seen in high-street chains. Younger designers in America still drew upon the country's history for leadership in sportswear, but names including Proenza Schouler and Rodarte translated these styles into luxurious forms, decorated with couture-inspired detailing. In London, new designers such as Todd Lynn, Louise Goldin, and Christopher Kane showed a revived interest in fine tailoring, inventive and colourful knitwear, and seasonally changing silhouettes, respectively.

Other cities across the globe were equally keen to tap into fashion as an exciting visual and material form. This was perhaps most clearly seen in India, China, South America, and the Pacific Rim where fashion weeks began to promote indigenous designers and seek new national and international names. In China, investments in production capacities were superseded by interest in design education and promotional trades, in order to build towards a strong fashion design profile for the future. In India and Russia, rising middle and upper-middle classes meant that a new cadre of people were keen to express their status and taste through clothing. New fashion magazines sprang up, both national editions of fashion staples such as *Vogue*, *Elle*, and *Marie Claire*, but also new titles that were inspired by local styles.

At street level, fashion was ever more evident, catalogued on websites such as http://www.thesartorialist.com, as well as sites that focused on the style of people in particular cities, from Stockholm to Sydney. These demonstrated fashion's continued ability to express individuality in permutations of existing fashions and emerging youth fashions. Subcultural fashion was equally vibrant, including reinventions of 1980s goth styles that spread internationally, and the allied teenage emo fashions. Club fashions were increasingly flamboyant, and referenced 1980s New Romanticism and Rave. As ever, fashion drew on its own history in order to move forward. It cross-referenced its past, and brought together new configurations of style. Thus, Christopher Kane was inspired by Azzedine Alaia's 1980s figure-hugging dresses and Versace's early 1990s vibrancy, but produced fashions that were new and fresh. New Rave reinvented its predecessor's neon colours and oversized slogan T-shirts. In each case, the new century saw an interest in volume and colour, which had been missing in much 1990s fashion.

The early 21st century also witnessed a growing number of ethically inspired labels and websites, which focused on fashion's impact on the planet as well as concern about workers' rights. This represented an important response to reports of exploitation in factories from Mexico to India, where garments were made for big Western brands. Fashion's need to address its production methods was a significant shift. While there had been calls for this since the mid-19th century, responses had been intermittent. It remains to be seen whether this boom in ethical fashion can infiltrate the industry as a whole and make permanent, far-reaching changes to the way textiles are made and clothing produced. It is to be hoped that this is a long-term trend, and not just a brief fashion.

Fashion had simultaneously grown as a subject of academic study, with increasing numbers of books and journals produced to examine its nature, status, and meaning. International museums presented fashion exhibitions to great acclaim and enormous fashion interest. At the other end of the market, the rise of celebrity culture spread fashions more quickly than even Hollywood had in its heyday. Cumulatively, these varied aspects of social, cultural, and political lifestyles and attitudes connected to the birth and dissemination of fashion, and its increasingly globalized character.

Fashion had not ended, therefore, but it had altered, and it was, potentially, on the brink of another major shift. As non-Western fashion systems grew in confidence, and recession set in, power could potentially shift towards the East. While it is unlikely that the Western fashion industry, which has evolved since the Renaissance, will be subsumed, it will have to adapt quickly to respond effectively to the global challenge.

References

Introduction

Arnold, Janet, *Patterns of Fashion: The Cut and Construction of Clothes* (Drama Book Publishers, 2008).

Arnold, Rebecca, *The American Look: Fashion, Sportswear and the Image of Women in 1930s and 1940s New York* (I. B. Tauris, 2009).

Barthes, Roland, *The Language of Fashion* (Berg, 2006).

—— *The Fashion System* (Jonathan Cape, 1985).

Breward, Christopher, *The Hidden Consumer: Masculinities, Fashion and City Life, 1860–1914* (Manchester University Press, 1999).

Entwistle, Joanne, *The Fashioned Body: Fashion, Dress and Modern Social Theory* (Polity Press, 2000).

Evans, Caroline, *Fashion at the Edge: Spectacle, Modernity and Deathliness* (Yale, 2007).

Hollander, Anne, *Seeing through Clothes* (University of California Press, 1993).

Lemire, Beverly, *Dress, Culture and Commerce* (Palgrave Macmillan, 1997).

Miller, Danny and Mukulika Banerjee, *The Sari* (Berg, 2003).

Ribeiro, Aileen, *Fashion and Fiction: Dress in Art and Literature in Stuart Britain* (Yale, 2005).

Wilson, Elizabeth, *Adorned in Dreams: Fashion and Modernity* (I. B. Tauris, 2003).

Chapter 1: Designers

Baillén, Claude, *Chanel Solitaire* (Collins, 1973), p. 69.
Beaton, Cecil, *The Glass of Fashion* (Cassell, 1989), p. 8.
Carter, Ernestine, 'Gabrielle "Coco" Chanel, 1883–1971: Magic of Self', in *Magic Names of Fashion* (Weidenfeld and Nicholson, 1980), pp. 52–66.
Dickens, Charles, *All the Year Round* (London, February 1863), quoted in Elizabeth Ann Coleman, *The Opulent Era: Fashion of Worth, Doucet and Pingat* (Thames and Hudson, 1989), p. 15.
Frankel, Susannah, *Visionaries: Interviews with Designers* (V&A Publications, 2001), pp. 34–5.
Rawsthorn, Alice, *Yves Saint Laurent: A Biography* (HarperCollins, 1996), p. 90.
Settle, Alison, *Clothes Line* (Methuen, 1937), p. 4.
Tetart-Vittu, Françoise, 'The French-English Go-Between: "*Le Modèle de Paris*" or the Beginning of the Designer, 1820–1880', in *Costume*, no. 26 (1992), pp. 40–5.
Williams, Beryl, *Young Faces in Fashion* (J. B. Lippincott, 1956), p. 145.

Chapter 2: Art

Apraxine, Pierre and Xavier Demarge, *'La Divine Comtesse': Photographs of the Comtesse de Castiglione* (Yale University Press, 2000), p. 13.
Hollander, Anne, *Seeing through Clothes* (University of California Press, 1993), p. xi.
Oliphant, Margaret, *Dress* (London, 1878), p. 4.
Ribeiro, Aileen, 'Fashion and Whistler', in Margaret F. MacDonald, Susan Grace Galassi, and Aileen Ribeiro, *Whistler, Women and Fashion* (The Frick Collection and Yale University Press, 2003), p. 19.
Stepanova, Vavara, 'Tasks of the Artist in Textile Production', in S. Novoer (ed.), *The Future is Our Only Goal*, p. 191, quoted in Radu Stern, *Against Fashion: Clothing as Art, 1850–1930* (MIT Press, 2004), p. 55.
Swanson, Carl, 'The Prada Armada', *New York Times Magazine* (16 April 2006).

Troy, Nancy, *Couture Culture: A Study in Modern Art and Fashion* (MIT Press, 2003), p. 7.

Warhol, Andy, *The Philosophy of Andy Warhol (From A to B and Back Again)* (Harcourt Brace and Company, 1977), p. 92.

Wollen, Peter, 'Addressing the Century', in Peter Wollen (ed.), *Addressing the Century: 100 Years of Art and Fashion* (Hayward Gallery Publishing, 1998), p. 16.

Chapter 3: Industry

Agins, Teri, *The End of Fashion: The Mass Marketing of the Clothing Business* (William Morrow, 1999), p. 5.

Godley, Andrew, 'The Emergence of Mass Production in the UK Clothing Industry', pp. 8–25, in I. Taplin and J. Winterton (eds.), *Restructuring in a Labour Intensive Industry: The UK Clothing Industry in Transition* (Avebury, 1996), p. 12.

Godley, Andrew, Anne Kershen, and Raphael Schapiro, 'Fashion and its Impact on the Economic Development of London's East End Womenswear Industry, 1929–62: The Case of Ellis and Goldstein', *Textile History*, vol. 34, no. 2 (November 2003), pp. 214–20.

Kidwell, Claudia and Margaret C. Christman, *Suiting Everyone: The Democratization of Clothing in America* (Smithsonian Institution, 1974), p. 39.

Lemire, Beverley, *Dress, Commerce and Culture: The English Clothing Trade before the Factory, 1660–1800* (Macmillan, 1997), pp. 122–4.

Moses, Elias, *The Growth of an Important Branch of British Industry: The Ready-Made Clothing System* (London: 1860), pp. 4–5.

Perrot, Philippe, *Fashioning the Bourgeoisie: A History of Clothing in the Nineteenth Century* (Princeton University Press, 1994), p. 54.

Chapter 4: Shopping

http://www.doverstreetmarket.com

Benson, Susan Porter, *Counter Cultures: Saleswomen, Managers, and Customers in American Department Stores, 1890–1940* (University of Illinois Press, 1988), p. 76.

Collins, Kenneth, speaking to the Fashion Group, New York, 13 September 1938, Box 73, File 2, Fashion Group Archives, New York Public Library.

Quant, Mary, *Quant by Quant* (Cassell, 1966), p. 43.

Rappaport, Erika, *Shopping for Pleasure: Women in the Making of London's West End* (Princeton University Press, 2000), p. 5.

Roche, Daniel, *A History of Everyday Things: The Birth of Consumption in France, 1600–1800* (Cambridge University Press, 2000), p. 213.

Smith, Woodruff D., *Consumption and the Making of Respectability, 1600–1800* (Routledge, 2002), pp. 44–51.

Thomas, Dana, *Deluxe: How Luxury Lost its Lustre* (Allen Lane, 2007), p. 300.

Chapter 5: Ethics

Arletty, quoted in 'Pour ou Contre les Pantalonnées', *L'Œuvre* (7 February 1942), cited in Dominique Veillon, *Fashion under the Occupation* (Berg, 2002), p. 127.

Cho, Margaret, quoted in Michael Slezak, 'Margaret Cho's Not Laughing About Gwen's Harajuku Girls', *Entertainment Weekly*, http://www.ew.com (2 November 2005).

Dunn, Jourdan, quoted in Elizabeth Day, 'How Racism Stalked the London Catwalk', *The Observer* (17 February 2008).

Jonson, Ben, *Epicoene or The Silent Woman*, ed. Roger Holdsworth (A&C Black, 1999), p. 100.

Killerby, Catherine Kovesi, *Sumptuary Law in Italy, 1200–1500* (Clarendon Press, 2002), p. 7.

Limnander, Armand, 'Slow Fashion', *The New York Times* (16 September 2007).

McDougall, Dan, 'The Hidden Face of Primark Fashion', *The Observer* (22 October 2008).

Spectator 49, quoted in Erin Mackie, *Market à la Mode: Fashion, Commodity and Gender in The Tatler and The Spectator* (John Hopkins University Press, 1997), p. 174.

Stitzel, Judd, *Fashioning Socialism: Clothing, Politics and Consumer Culture in East Germany* (Berg, 2005), p. 3.

Wolf, Jaime, 'And You Thought Abercrombie and Fitch Was Pushing It?', *The New York Times* (23 April 2006).

Chapter 6: Globalization

Armstrong, Lisa, 'A Little Local Colour Goes a Long Way', *The Times* (16 February 2006).

Asome, Caroline and Alan Hamilton, 'Former Duckling Grows into Swan of Global Fashion', *The Times* (15 February 2005).

Belfanti, Carlo Marco, 'Was Fashion a European Invention?', in *Journal of Global History*, no. 3 (2008), pp. 419–43.

Jones, Ann Rosalind and Peter Stallybrass, *Renaissance Clothing and the Materials of Memory* (Cambridge University Press, 2001), p. 57.

Maynard, Margaret, *Dress and Globalisation* (Manchester University Press, 2004), pp. 2–5.

Miyake, Issey, quoted in Mark Holborn, *Issey Miyake* (Taschen, 1995), p. 16.

Mustafa, Hudita Nina, 'La Mode Dakaroise: Elegance, Transnationalism and an African Fashion Capital', in Christopher Breward and David Gilbert (eds.), *Fashion's World Cities* (Berg, 2006), pp. 177–200.

Nagrath, Sumati, 'Local Roots of Global Ambitions: A Look at the Role of the India Fashion Week in the Development of the Indian Fashion Industry', in Jan Brand and José Teunissen (eds.), *Global Fashion, Local Tradition: On the Globalisation of Fashion* (Terra, 2005), p. 49.

Neissen, Sandra, 'The Prism of Fashion: Temptation, Resistance and Trade', in Jan Brand and José Teunissen (eds.), *Global Fashion, Local Tradition: On the Globalisation of Fashion* (Terra, 2005), p. 165.

Rabine, Leslie W., *The Global Circulation of African Fashion* (Berg, 2002), p. 3.

Ribeiro, Aileen, *Dress in Eighteenth Century Europe, 1715–1789* (Batsford, 1984), pp. 169–70.

Sato, Tomoko and Toshio Watanabe, 'The Aesthetic Dialogue Examined: Japan and Britain, 1850–1930', in Tomoko Sato and Toshio Watanabe (eds.), *Japan and Britain: An Aesthetic Dialogue, 1850–1930* (Lund Humphries, 1991), pp. 38–40.

Styles, John, 'Tudor and Stuart Britain, 1500–1714: What was New?', in Michael Snodin and John Styles (eds.), *Design and the Decorative Arts: Britain 1500–1900* (V&A Publications, 2001), p. 136.

Teunissen, José, 'Global Fashion/Local Tradition: On the Globalisation of Fashion', in Jan Brand and José Teunissen (eds.), *Global Fashion/Local Tradition: On the Globalisation of Fashion* (Terra, 2005), p. 11.

Vicente, Marta V., *Clothing the Spanish Empire: Families and the Calico Trade in the Early Modern Atlantic World* (Palgrave Macmillan, 2006), pp. 65–6.

Further reading

Introduction

Breward, Christopher, *The Culture of Fashion: A New History of Fashionable Dress* (Manchester University Press, 1995).

Bruzzi, Stella and Pamela Church Gibson, *Fashion Cultures: Theories, Explorations and Analysis* (Routledge, 2000).

Jarvis, Anthea, *Methodology* Special Issue, *Fashion Theory: The Journal of Dress, Body and Culture*, vol. 2, issue 4 (November 1998).

Kawamura, Yuniya, *Fashion-ology: An Introduction to Fashion Studies* (Berg, 2004).

Kaiser, Susan, *Social Psychology of Clothing: Symbolic Appearances in Context* (Fairchild, 2002).

Purdy, Daniel Leonhard (ed.), *The Rise of Fashion: A Reader* (University of Minnesota Press, 2004).

Chapter 1: Designers

Aoiki, Shoichi, *Fresh Fruits* (Phaidon, 2005).

Kawamura, Yuniya, *The Japanese Revolution in Paris Fashion* (Berg, 2004).

Muggleton, David, *Inside Subculture: The Postmodern Meaning of Style* (Berg, 2000).

Seeling, Charlotte, *Fashion: The Century of Designers, 1900–1999* (Konemann, 2000).

Steele, Valerie and John Major, *China Chic: East Meets West* (Yale, 1999).

Chapter 2: Art

Francis, Mark and Margery King, *The Warhol Look: Glamour, Style, Fashion* (Little, Brown and Company, 1997).

Martin, Richard, *Fashion and Surrealism* (Thames and Hudson, 1989).

Radford, Robert, 'Dangerous Liaisons: Art, Fashion and Individualism', in *Fashion Theory: The Journal of Dress, Body and Culture*, vol. 2, issue 2 (June 1998).

Ribeiro, Aileen, *The Art of Dress: Fashion in England and France, 1750–1820* (Yale, 1995).

Townsend, Chris, *Rapture: Art's Seduction by Fashion since 1970* (Thames and Hudson, 2002).

Winkel, Marieke de, *Fashion and Fancy: Dress and Meaning in Rembrandt's Painting* (Amsterdam University Press, 2006).

Chapter 3: Industry

Gereffi, Gary, David Spencer, and Jennifer Bair (eds.), *Free Trade and Uneven Development: The North American Apparel Industry after NAFTA* (Temple University Press, 2002).

Green, Nancy, *Ready-to-Wear, Ready-to-Work: A Century of Industry and Immigrants in Paris and New York* (Duke University Press, 1997).

Jobling, Paul, *Fashion Spreads: Word and Image in Fashion Photography since 1980* (Berg, 1999).

McRobbie, Angela, *British Fashion Design: Rag Trade or Image Industry?* (Routledge, 1998).

Phizacklea, Annie, *Unpacking the Fashion Industry: Gender, Racism and Class in Production* (Routledge, 1990).

Tulloch, Carol (ed.), *Fashion Photography*, Special Edition of *Fashion Theory: Journal of Dress, Body and Culture*, vol. 6, issue 1 (February 2002).

Chapter 4: Shopping

Benson, John and Laura Ugolini, *Cultures of Selling: Perspectives on Consumption and Society since 1700* (Ashgate, 2006).

Berg, Maxine and Helen Clifford (eds.), *Consumers and Luxury: Consumer Culture in Europe, 1650–1850* (Manchester University Press, 1999).

Lancaster, Bill, *The Department Store: A Social History* (Leicester University Press, 1995).

Leach, William, *Land of Desire: Merchants, Power and the Rise of a New American Culture* (Vintage, 1993).

Richardson, Catherine (ed.), *Clothing Culture, 1350–1650* (Ashgate, 2004).

Shields, Rob, *Lifestyle Shopping: The Subject of Consumption* (Routledge, 1992).

Worth, Rachel, *Fashion for the People: A History of Clothing at Marks and Spencer* (Berg, 2007).

Chapter 5: Ethics

Arnold, Rebecca, *Fashion, Desire and Anxiety: Image and Morality in the Twentieth Century* (I. B. Tauris, 2001).

Black, Sandy, *Eco-Chic: The Fashion Paradox* (Black Dog, 2008).

Guenther, Irene, *Nazi Chic: Fashioning Women in the Third Reich* (Berg, 2005).

Ribeiro, Aileen, *Dress and Morality* (Berg, 2003).

Ross, Andrew (ed.), *No Sweat: Fashion, Free Trade and the Rights of Garment Workers* (Verso, 1997).

Chapter 6: Globalization

Bhachu, Parminder, *Dangerous Designs: Asian Women Fashion the Diaspora Economies* (Routledge, 2004).

Clark, Hazel and Eugenia Paulicelli (eds.), *The Fabric of Cultures: Fashion, Identity, Globalization* (Routledge, 2008).

Eicher, Joanne B. (ed.), *Dress and Ethnicity: Change across Space and Time* (Berg, 1999).

Kuchler, Susanne and Danny Miller, *Clothing as Material Culture* (Berg, 2005).

Niessen, Sandra, Ann Marie Leshkowich, and Carla Jones (eds.), *Re-Orienting Fashion* (Berg, 2003).

"牛津通识读本"已出书目

古典哲学的趣味	福柯	地球
人生的意义	缤纷的语言学	记忆
文学理论入门	达达和超现实主义	法律
大众经济学	佛学概论	中国文学
历史之源	维特根斯坦与哲学	托克维尔
设计,无处不在	科学哲学	休谟
生活中的心理学	印度哲学祛魅	分子
政治的历史与边界	克尔凯郭尔	法国大革命
哲学的思与惑	科学革命	丝绸之路
资本主义	广告	民族主义
美国总统制	数学	科幻作品
海德格尔	叔本华	罗素
我们时代的伦理学	笛卡尔	美国政党与选举
卡夫卡是谁	基督教神学	美国最高法院
考古学的过去与未来	犹太人与犹太教	纪录片
天文学简史	现代日本	大萧条与罗斯福新政
社会学的意识	罗兰·巴特	领导力
康德	马基雅维里	无神论
尼采	全球经济史	罗马共和国
亚里士多德的世界	进化	美国国会
西方艺术新论	性存在	民主
全球化面面观	量子理论	英格兰文学
简明逻辑学	牛顿新传	现代主义
法哲学:价值与事实	国际移民	网络
政治哲学与幸福根基	哈贝马斯	自闭症
选择理论	医学伦理	德里达
后殖民主义与世界格局	黑格尔	浪漫主义

| 批判理论 | 德国文学 | 儿童心理学 |
| 电影 | 戏剧 | 时装 |